WATER IS...

The Indispensability
of Water in Society
and Life

With a Foreword by
Seth M. Siegel, Author of *New York Times*
bestseller *Let There Be Water*

WATER IS

The Indispensability
of Water in Society
●●● and Life

Seth B. Darling

Seth W. Snyder

World Scientific

NEW JERSEY · LONDON · SINGAPORE · BEIJING · SHANGHAI · HONG KONG · TAIPEI · CHENNAI · TOKYO

Published by

World Scientific Publishing Co. Pte. Ltd.

5 Toh Tuck Link, Singapore 596224

USA office: 27 Warren Street, Suite 401-402, Hackensack, NJ 07601

UK office: 57 Shelton Street, Covent Garden, London WC2H 9HE

Library of Congress Cataloging-in-Publication Data
Names: Darling, Seth B., author | Snyder, Seth W., author.
Title: Water is... : the indispensability of water in society and life / Seth B. Darling
    (University of Chicago, USA), Seth W. Snyder (Northwestern University, USA).
Description: New Jersey : World Scientific, 2018.
Identifiers: LCCN 2018018584 | ISBN 9789813271395 (hardback : alk. paper)
Subjects: LCSH: Water and civilization.
Classification: LCC CB482 .D37 2018 | DDC 333.91--dc23
LC record available at https://lccn.loc.gov/2018018584

British Library Cataloguing-in-Publication Data
A catalogue record for this book is available from the British Library.

First published 2018 (Hardback)
Reprinted 2018 (in paperback edition)
ISBN 978-981-3278-10-3 (pbk)

For any available supplementary material, please visit
https://www.worldscientific.com/worldscibooks/10.1142/11018#t=suppl

Desk Editor: Amanda Yun

Typeset by Stallion Press
Email: enquiries@stallionpress.com

For Adina, my beloved, beautiful collection of mostly water.

— SBD

In memory of Irwin N. Snyder, who taught me that water
is the solution to putting out fires and saving lives.

— SWS

Foreword to *Water Is*

By **Seth M. Siegel,** Author of *The New York Times*
bestseller *Let There Be Water*

After a short-ish career as a lawyer and a long-ish career as an entrepreneur and business executive, for still unexplained reasons, I became infatuated with water. I wanted to know everything I could know. I read widely, but randomly. Everything was on the list: Water policy, governance, irrigation, pricing, history, biology, pollution, vocabulary, turbidity, appliances, desalination, rationing, wastewater treatment, and on and on. Whatever popped into my mind became the topic of the day or week, and lots of underlined books and articles started piling up on my desk.

While that kind of footloose and fancy free approach to learning was fun — allowing myself to be carried wherever the whim took me — it left me open to putting too much emphasis on less important ideas and, sometimes, missing out on what was a truly major area of water. If there is a hydrological version of the saying "missing the forest for the trees," please insert it here.

What I didn't have was a syllabus or reading list. Until now.

Thanks to the quirk of Seth Darling and Seth Snyder and I sharing a fairly unusual first name — and all three of us attending an event to mark the announcement of an endowed chair at the University of Chicago (where the other Seths have had professional associations as well as at US Department of Energy research laboratories) — we began talking and realized we three shared a passion. The passion wasn't just for water, as you'd imagine. It was also for sharing the meaning of water with anyone who cared to listen and learn.

One of the Seths mentioned that they were working on a book together about the relevance of water, and I told them I'd be excited to read it. One of them asked if I would consider writing the Foreword for the book. I didn't hesitate a moment to say Yes, and then one of us suggested the title of the book being "Three Seths on Water." (I don't remember who suggested it, but I hope it was me because I think it was a good joke — at least as far as academic or water jokes go.)

Time passed, and I mostly fell out of touch with the two scientists. But then a typed manuscript arrived one Friday, and before the weekend was over, I had devoured **Water Is**.

I had several reactions.

First was: These guys are good writers. They took complex material and, with wit and style, boiled it down so that any educated reader could enjoy the book, and learn a great deal in a short time.

Second was: How smart of them to structure the book as they did. Everyone says how water is connected to everything. But they didn't just say it. They proved it. With separate chapters on the science, sociology, governance, culture and economy of water that often referenced other parts of the book, they brought it all together.

And third was: A wish that **Water Is** had been written when my interest in water began and morphed into infatuation and then into my life's work. I'd have saved so much time running around and would have compactly learned so much, so quickly — and with the right emphasis.

With world population rising and affluence growing, we have a big problem on the horizon. We are now about seven billion in the world, projected to rise to more than ten billion by mid-century. Further, it is very good news that people are rising out of poverty, but as they do, their diets become much more based on animal protein, a food source that is orders of magnitude more water consumptive than plant-based nutrition. Add to these two mega-trends that climate change is drying up water sources due to faster evaporation and less rainfall where and when it is needed.

These three present a perfect storm: Greater demand and less supply for a commodity that is already often in short supply. What is needed, and needed urgently, is more widespread awareness of coming water woes and ideas on what to do about it.

As *Water Is* is a one-of-a-kind book, let's hope that the shelves of our bookstores become crammed with other volumes taking on other aspects of the issue. If fifty years ago, there wasn't a single US university that offered a course, let alone a major, in environmental studies, the rise of the environmental movement has it that today it is unlikely that there is any significant school of higher learning without a whole department focused on it. Perhaps some years from now, there will be departments of Water Studies that incorporate — as does *Water Is* — chemistry, physics and biology, but also history, public policy and economics.

One idea that doesn't need to await the armies of PhDs who will be getting their degrees in Water Studies is to begin charging the real price for water. As Seth and Seth point out, water charges are often based on the delivery of the water to the home or business, and not on the cost of the product itself. Imagine if Amazon announced that going forward all products ordered would be free, and only a small shipping charge would be billed. People would "buy" all kinds of things they don't need. Related to water, at "free," the laws of economics disappear and no one is incentivized to conserve or to search for technologies that could either minimize the use of water or, for that matter, to invent those technologies.

And in many places, even the water charges, as minimal as they are today for that delivery, essentially go away. In much of the world, water is provided as a public good, a benefit used by governments to buy the support of farmers and city dwellers alike. Akin to that, in many unmetered places, there is a flat monthly or annual fee for water. Systems like these provide incentives to "get your money's worth," and, as a result, to leave the spigot on overnight. If not that cavalier about wasting water, for sure, no home owner would have a reason to call in the plumber and pay to fix a leak or for any farmer to switch from flood irrigation to some form of micro-irrigation that they would have to buy like drip irrigation.

If there is a single variable that needs changing in how we think about water, it is the pricing of it. If the full price were charged, a virtuous cycle would begin almost immediately. There would be more money for critical infrastructure and the several hundred thousand annual water main breaks in the US alone would stop disrupting traffic and wasting water.

The creative genius of the world's inventors would be unleashed, coming up with as many ways to save water as there are ways to lose or to waste it. If one dollar of technology can save two dollars on water bills, everyone will be buying that technology.

With more people focused on water, so, too, will our elected officials be. Now, water is mostly found in the "out of sight, out of mind" category because few citizens complain about aged wastewater treatment or other water needs until they become a front-page crisis.

And our universities will be developing more and more courses to educate students eager to be hired at the ever-greater number of water companies hiring the best and the brightest. A true theme for freshman orientation speeches could then be: Want to have a satisfying career and save the world? Go to work in the world of water.

Or we can do little but putter along, doing, more or less, more of what we've been doing. But if we do, no one should be surprised if a water problem here and there grows worse, ultimately morphing into a humanitarian tsunami. Around the world, hundreds of millions of people are relying on water that is at risk of giving out. If that does, food prices will rise, social instability will grow, countries that are already economically wobbly will be at risk of becoming failed states, and even large flows of refugees may be forced to pick up from where they live in search of some welcoming place.

Among the many wonderful elements of **Water Is** is a fascinating section about water in antiquity. In every case, Seth and Seth show, the ancient civilizations that thrived developed new water technologies for farming and/or bringing water to their cities. It made their people healthy thanks to good nutrition and their societies economically powerful. That health and wealth gave them the ability to dominate their regions, and sometimes areas beyond that. And in each case that the water gave out because of climate change or salinity of their water or some other cause, that great civilization soon fell.

Reading about these ancient peoples, it is hard to see it only as a history lesson. Learning about once-successful nations that failed because of water woes is a reminder to us of the illusion of permanence and the lesson that their fate can be ours, too.

But none of that has to be the case. If our future is in the hands of Seth and Seth and people inspired by them, we can innovate our way out of nearly any challenge long before it becomes an enduring problem.

Introduction: Three Seths on Water

We are Seth Darling and Seth Snyder, two physical chemists who became colleagues at the U.S. Department of Energy's Argonne National Laboratory. [Here we are obliged to tell you that this book represents the views of the authors and not those of our employer(s).] Argonne is located in the suburbs of Chicago. We Seths have lived a significant fraction of our adult lives near beautiful Lake Michigan. How do two scientists who work at an energy laboratory and live a stone's throw from one of the most water-secure places in the world develop a passion for water? We hope *Water Is* tells that story. It's a journey that will travel through time, touching down here and there through the history of the world, weaving connections to our present society, and providing a glimpse into our shared water future. The science of water is unique, endlessly surprising, and deeply fascinating — a message we hope you'll share after reading this book. Water is so common, it is easy to overlook just how weird and wonderful it is. The engineering of water is an epic tale of human achievement against (or with) the elements. The role of water in our lives, culture, and world is second to none. As our research careers progressed, individually we both became aware of the central role that water plays specifically in energy. We need energy to provide water and we need water to provide energy. Frequently one is the largest barrier to the other.

Our production and use of energy is a primary driver of greenhouse gas emissions. Energy use propels climate change, but many of the impacts are manifested through water — from parching droughts to relentless sea level rise. As energy scientists, our focus on water is an acknowledgment that, despite increasing attention and burgeoning collective efforts, society will not progress fast enough to curb greenhouse gas emissions and avoid major impacts of climate disruption. If we cannot avoid these changes, we must identify and attempt to mitigate the impacts. Society has developed solutions to managing water going back to the dawn of civilization. These technologies and social solutions entirely addressed local challenges. We are now faced with global challenges, but in many ways we can follow the strategy of our ancestors. If we develop innovative local or regional solutions, we can start addressing the global challenges that have only emerged in the authors' lifetimes.

One way to think about water is the trifecta of quantity, quality, and price — generally speaking, you can pick two. If we want water to be plentiful and high quality, then we must be ready to pay for it. The challenge is that most societies consider clean, plentiful water a fundamental right, not a resource only for the wealthy. From that perspective, society has two routes: develop solutions to reduce costs of providing usable water or subsidize water for those who cannot afford it. You can guess which of these the authors believe the wiser option.

So that's a story of two Seths. But why stop at two when you can have three? Seth Siegel is a *New York Times* bestselling author and globally recognized expert on water innovation and policy. When we were considering who might be a good fit to write a foreword for this work, it was hard to imagine a better person than Seth. He has deep insight on the subject and knows well how to communicate these complex issues with the general public. And his name is Seth. C'mon, you have to admit that's pretty cool.

There are hundreds of books written on water, dozens of journals that report results of water research, and many non-profits that focus on water. Each comes at the multifaceted, enthralling, and at times alarming subject of water from its own perspective. We hope our work contributes to this topic a concise, contemporary, and informative view of what *Water Is*.

Contents

01 Water is Chemistry and Physics

Origins of water

Viewed from space, our planet's most remarkable feature is its water. It covers around 70% of Earth's surface, and participates in virtually all meteorological, geological, chemical, and biological processes. Most of us learned about the water cycle in school, but rarely is the question "Where did all the water come from in the first place?" posed. You might think that is a simple question with an established answer. This is not the case. The origins of Earth's water remain a subject of intense study and debate in the scientific community. Let's start by going back even further in time — all the way back to the Big Bang. In the moments following the birth of the Universe, primordial particles joined to form considerable amounts of hydrogen, by far the most plentiful element even today. Hydrogen is one of the ingredients that make up water, as we all know from the world's most famous chemical formula: H_2O. Oxygen, the other ingredient, was not created directly from the Big Bang. To find oxygen we have to wind the clock forward a billion or so years to a time when stars populated the universe. Inside the dense, intensely hot cores of stars, light elements like hydrogen and helium fuse together into heavier ones, including oxygen. When a massive star explodes in a supernova, the colossal blast spews these fused elements out into space. When hydrogen and oxygen meet up in interstellar clouds, they merge to form water molecules. So that is, in a sense, where our water came

■1

from, and that part is not under dispute. But how did it get from space into our oceans and lakes? Here is where things get less certain.

Our solar system, which formed about nine billion years after the Big Bang, condensed from one such dusty interstellar cloud that certainly contained water. It is reasonable to conclude, then, that Earth's water has always been on Earth, being a constituent of the materials that originally formed the planet. However, in Earth's early history, there was not a well-formed atmosphere as we know it today and the surface temperatures were scalding. Any water existing on the surface at that time would most likely have evaporated and escaped into space, so the water we see on Earth today must have been delivered to the surface much later. We can look up for a possible answer: comets and asteroids. Both of these extraterrestrial bodies contain water ice, and many collided with Earth (especially in the young solar system), thereby depositing water onto the planet. As to whether it was asteroids or comets that contributed a greater share, scientists continue to deliberate on that topic. Do comets and/or asteroids answer the question, then? Not necessarily. It is indeed likely that at least some of the water on Earth came via bombardment from space, but there is another option to consider. We can instead look down for a possible answer: deep water. It may be that a vast amount of the water that was present when the Earth formed remained safely trapped inside the planet during those atmosphere-less eons. Earth has an active core that, over geologic time scales, can tectonically exchange materials between the crust and the interior. Perhaps this slow exchange brought water to the surface to form oceans after the planet had cooled sufficiently to maintain a stable atmosphere. Somewhat surprisingly, it is often easier for scientists to probe the chemical makeup of a distant planet than that of our own planetary core. This is because scientists use light and other radiation signals to understand chemical makeup. Such signals can travel through light years of empty space but won't penetrate from the depths of the Earth's interior. However, once in a while the deep mantle that is otherwise inaccessible delivers a gift to researchers in the form of rocks expelled through volcanic activity. By studying these specimens and probing subsurface structure through analysis of seismic waves, scientists have found evidence for a shockingly large amount of water buried deep below the surface in a transition zone between the upper and lower mantle — perhaps even several times the total volume of water that we see in our oceans. This water is not free like the water on the surface, but rather it is locked up in the rock itself. No doubt there remains much to learn about the origins of Earth's water, but water's mysteries extend beyond just its beginnings.

An anomalous liquid

We experience water in various forms every day of our lives. It is the most common and ubiquitous liquid. It is odorless and tasteless. What we smell and taste in water is actually from the presence of other species. Water is so mundane that it seems it could not be less remarkable. On the contrary, despite its apparent ordinariness, water is the most anomalous liquid we encounter. In *The Immense Journey*, anthropologist and author Loren Eiseley said "If there is magic on this planet, it is contained in water." Water is undeniably astonishingly bizarre and, surprisingly, still not fully understood. We will get to all sorts of things we do know about water shortly, some of which make clear just how weird water really is, but first let's take a moment to go over what we don't know. A single water molecule is not a complicated thing. It is comprised of one oxygen atom and two hydrogen atoms. The geometric arrangement of these atoms is of pivotal importance in determining the properties of water as a material, when lots of molecules come together. The three atoms in a water molecule are not arranged in a straight line, rather, it is bent with both hydrogen atoms sitting on one side of the oxygen atom (Figure 1.1). Many other common molecules form straight lines and do not possess the unique properties of water.

Each hydrogen atom is covalently bonded to the oxygen atom, consuming two of the six electrons available in oxygen's outer electron shell. The bent structure is a result of oxygen also having two so-called lone

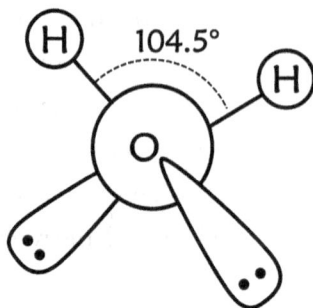

Figure 1.1.   Structure of a water molecule.

pairs of electrons that do not bond to other atoms. The bent structure and the lone pairs make water much more chemically active and interactive than other simple molecules. The covalent bonds and the lone pairs all repel each other, forcing the hydrogen atoms onto the far side of the molecule, forming a bond angle of 104.5°. This structure maximizes the distance between the hydrogen atoms and oxygen's electron pairs, with each item pointing to the corner of a virtual tetrahedron, a three-dimensional structure with four similar surfaces and corners.

It is the interactions between these simple water molecules that remains something of a mystery despite over a century of scientific study. There is a complex dance among groups of water molecules, with different molecules coming together and separating in extremely fast fluctuations. The structures of these transient clusters and dynamics of this process are maddeningly challenging to explore even using state-of-the-art characterization instruments and high-level theory. It is quite likely that hidden inside these ephemeral clusters are insights that could transform our understanding of biology, chemistry, geology, and more.

With that in mind, let's get back to things we understand pretty well when it comes to water. Water's bent shape has ramifications of importance that is difficult to overstate. Foremost among these is that, because oxygen favors electrons more so than does hydrogen (oxygen is more electronegative), the electron cloud surrounding the molecule is asymmetric. That is, the population of electrons on the oxygen side of

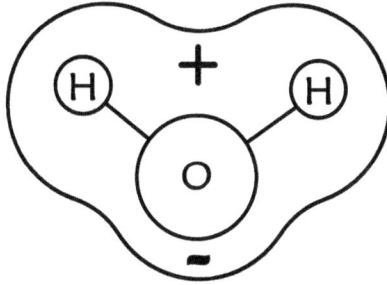

Figure 1.2.   Polarity of a water molecule.

the molecule is higher than that on the hydrogen side, leading to a net positive charge in the vicinity of the hydrogens (Figure 1.2) and a net negative charge near the oxygen. The term describing this phenomenon is "polarity," and it is arguably the most important feature of water.

Water is the only substance on Earth that is found naturally in three phases: solid, liquid, and gas. That in itself is a curious fact, but a closer look reveals an even more unusual observation regarding the states of matter for water. Polarity leads to a rare property of water, namely, that solid water (ice) is less dense than liquid water. Upon freezing, the density of water drops nearly 10%. You've surely noticed this yourself in that ice cubes float in your beverages. Almost all known materials have the reverse relation. It turns out that the maximum density for water actually occurs at 4 °C. (Under increasing pressure, ice will eventually undergo transitions to other crystalline structures with higher density, but these conditions do not typically exist on Earth — except in Kurt Vonnegut novels.) Ice has a lower density because the molecules arrange themselves into a stable crystalline structure with the polar species orienting to minimize contact between like charges. Ice's structure consumes a greater volume per unit molecule than those mysterious disordered water clusters present in liquid water. This strange feature is crucial for life. When bodies of water freeze in the winter, ice floats to the surface and provides an insulating barrier allowing the water below to remain liquid throughout the year, thereby sustaining aquatic life in cold climates. If ice did not float, then it would sediment and possibly

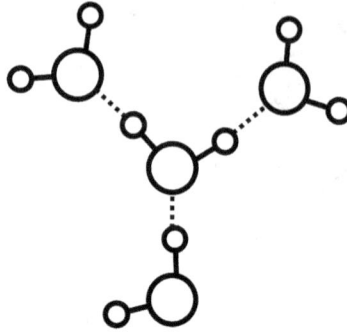

Figure 1.3.   Hydrogen bonding.

even remain frozen the next summer. Many bodies of water would slowly freeze over, and the planet would be uninhabitable.

The interaction between water molecules occurs primarily through so-called hydrogen bonding. A hydrogen bond is an attractive electrostatic force between a hydrogen atom bound to an electronegative atom (such as oxygen) and a second electronegative atom. The hydrogens in water fit this description, and this leads to an attraction between neighboring water molecules (Figure 1.3).

Hydrogen bonds are generally not as strong as covalent bonds, but that does not mean they are not significant. Water, in particular, exhibits rather robust hydrogen bonding. One consequence of water's hydrogen bonds is that the boiling point of water is higher than one might expect from its mass alone. Normally, lightweight molecules like water are gases at room temperature and pressure, but not water. Intermolecular hydrogen bonding makes it harder for the molecules to separate from one another and enter the gas phase; this allows water to exist in its liquid form on most of our planet. Another consequence of hydrogen bonding is that water is sticky. The technical terms for this stickiness are cohesion and adhesion, with the former describing water's attraction to water and the latter water's attraction to other substances. Water's cohesion is actually the highest among all non-metallic liquids, yet another example of its anomalous nature. Water forms drops so easily thanks to its cohesion. A correlated property is water's surface tension — also the highest among common liquids. Water molecules sitting at the surface do not have as many neighboring molecules as their brethren buried in the bulk, so these surface molecules cohere more strongly to their neighbors. The surface layer experiences a net inward force, pulled by the molecules below. Therefore, these surface molecules contract a bit, meaning the surface is under tension. Surface tension allows small insects like the water strider and even some amazing lizards to walk or run on water, provides the mechanism by which soap and detergents work (by lowering water's surface tension), and explains why bubbles are round.

Yet another effect of hydrogen bonding is a phenomenon called capillary action. When water is confined within a narrow tube or porous material, adhesion between water and the walls causes the liquid to rise along the edges. Surface tension acts to maintain an intact surface, resulting in a smooth, curved surface. If the adhesion is stronger than the water's cohesion, the water will lift upward into the tube or material, stopping only when gravity overcomes these forces. Capillary action is why paper towels wick up fluids so readily, and it allows plants to survive by carrying water through the xylem from the roots up into the limbs and leaves.

Strong hydrogen bonding in water also provides it with a specific heat capacity exceeding that of all common liquids and solids. Heat capacity is a measure of how much the temperature of a material changes upon the addition of a certain amount of heat. A material with a high heat capacity will experience relatively little change in temperature. Since water molecules are strongly attracted to one another, it takes more energy to make them move around. At the molecular level, temperature can be viewed as proportional to the average energy associated with molecular motion, so less motion means lower temperature. Water's anomalously high heat capacity makes it a good heat storage medium, coolant, and heat shield.

Another anomalous feature of water is its color. Yes, even pure water does have a color, albeit faint and, yes, it's blue. Bodies of water generally appear to be blue, of course, but that color is in part a product of impurities, in part a result of scattering of light by particulates suspended in the water and density fluctuations, in part reflection of the blue sky above (itself blue because of scattering), and in part because water itself is indeed blue. To see why water's color is anomalous, we first have to review the origin of color in most common materials. Light from the sun or typical light bulbs contains a full spectrum of colors, evident when it is dissected into a rainbow. When combined all together, these colors become white light. If a transparent material absorbs parts of that spectrum, say the reds, then only the remaining colors will pass through, making it appear blue. (Similarly, if an opaque material absorbs parts of the spectrum and reflects or scatters others, it will appear to be the color of the reflected or scattered light.) Color-generating absorption processes almost always involve the excitation of an electron from a lower energy state to a higher one, an electronic transition, with the color of light absorbed corresponding to the energy difference between these two states. Water, evidently, would not be satisfied with such an ordinary way of creating color. The states for electrons in water do not have levels separated with energies corresponding to visible colors. Vibrations within a water molecule itself, rather, are actually the origin of water's blue tinge. Oxygen-hydrogen bond vibrations have an energy that corresponds to light having a wavelength of about three microns — that's in the infrared part of the spectrum, invisible to the human eye. So it also is not simply fundamental excitation of these vibrations that explains the blue color either. It is excitation of higher level vibrations, something that is a rather unlikely event as light passes through water, which results in the blue tinge. These absorptions are so weak, in fact, that it was only in recent years that they have been measured accurately. That's why your glass of water looks clear and colorless; you need an awful lot of water, like in a large pool or the ocean, before the delicate blue is apparent.

The universal solvent

In addition to all of these extraordinary physical properties of water that result in large part from its polarity, water's polar nature also makes it a phenomenally effective solvent. That is, water is capable of dissolving a breathtaking variety of different substances; it is, in fact, often called the "universal solvent." Those materials that are attracted to water (and therefore are likely to dissolve well) are anthropomorphically called "hydrophilic" (water-loving) and those that are not are called "hydrophobic" (water-fearing). When ionic or polar molecules such as acids, salts, sugars, or alcohols encounter water, many water molecules surround them with the negatively charged oxygen sides of water attracting the positively charged components of the solute and the positively charged hydrogen sides attracting the negative components. Dissolved molecules are therefore said to be "hydrated" by a shell of water molecules. It is even possible to get small amounts of some non-polar substances such as carbon dioxide or gasoline to dissolve in water via a process where water either induces mild polarity in the solute or there is a gain in entropy (greater dispersal of the solute molecules relative to the pure materials) from the dissolution process. Since the Big Bang, nature strives to increase entropy. Water's unique ability to dissolve so many different substances is critical for life. Our bodies, like those of all living things, contain many thousands of different chemicals — each solvated in water. Wherever water goes, whether through the soil and rocks, through our bodies, or through the atmosphere, it grabs other substances and takes them along for the ride. Such action carries valuable minerals and nutrients to sustain life and reshape the world around us, but it has an unfortunate flip side — water is also extremely easy to pollute.

Let's take a look at some examples of the universal solvent in action and its central importance. One of the non-polar substances that can dissolve in water, although only in relatively low concentrations (just a few milligrams per liter), is oxygen (O_2). Many organisms living in the water depend on this dissolved oxygen (DO) to breathe. There is a delicate balance in aquatic environments where various dissolved substances can promote or hinder different species. Too many nutrients in the water from agricultural runoff or sewage can spur the growth of algal or phytoplankton blooms, which consume DO and create eutrophic conditions — dead zones. So how does oxygen get in the water in the first place? It comes right out of the air above the water. Oxygen prefers not to mix into water, but there is so much exposure from the atmosphere that a fraction does dissolve and form an equilibrium. The equilibrium level arises from a balance between the rates from the prevalent atmospheric oxygen molecules slowly mixing into the water and the release of low-concentration oxygen molecules dissolved in the water. Movement of the water increases interactions with the air and increases DO. Therefore, stagnant water tends to contain less DO and rapidly moving water more. Cold water contains more DO and warm water less. Both of these relations are driven primarily by the influence of photosynthesis (warm, still water supports more photosynthetic organisms, which consume oxygen). Hot, calm weather can even reduce DO in lakes and ponds sufficiently to kill off large numbers of fish.

Ionic (charged) species, unlike such non-polar substances, tend to be highly soluble in water. Calcium and magnesium carbonates are a common example, which are the key culprits in so-called hard water. Many aquifers, which are tapped with wells for human use, are located in carbonate caverns that contribute ions into the groundwater. Minerals such as these dissolved in water are usually not a health concern, but they can be a nuisance. Hard water decreases the effectiveness of soaps and detergents, increasing soap scum deposits. It can shorten the life of fabrics. Hard water will leave minerals behind in your coffee maker (which you can later dissolve using a mild acid like vinegar). Long-term movement of hard water through pipes and equipment can result in buildup of scale on the walls, decreasing pressure and damaging appliances like water heaters. To mitigate these troubles water is often softened. In a residential setting, softening is often accomplished with ion exchange resins that swap out the calcium and magnesium for sodium or potassium. Sodium and potassium ions are more soluble than calcium and magnesium, and will only form scale at significantly higher concentrations. Note that the total dissolved solids (TDS) are not changed by this process, but the replacement ions have fewer negative impacts on plumbing and the like. Another strategy is lime softening, or Clark's process, where calcium hydroxide is added to precipitate out the carbonates — in this case actually lowering the TDS in the water. Other, less common approaches involve chelation, distillation, or reverse osmosis membrane filtration (see Chapters 6 and 8 for more on these purification methods).

Hard water does contain salt, but the dissolved salts in household water would rarely exceed 200 parts per million (ppm), which is well within the range of fresh water. The parts per million refers to the ratio of salt molecules to total molecules (water plus salt). Water is not considered saline until the concentration exceeds 1,000 ppm (see Table 1.1).

In the oceans, the primary contribution to the salinity comes from good ol' sodium chloride (table salt, existing as Na^+ and Cl^- ions in water), with far lower amounts of magnesium, calcium, potassium, sulfate, and other ions. When salt is dissolved in water, the density of the water increases. This is why things are more buoyant in salty water than in fresh water, exemplified by briny bodies of water like the Dead Sea, which is about ten times saltier than typical seawater. Saline water also has a lower freezing point than fresh water (about 1.9°C lower for seawater). This is why roads are salted in the winter to limit ice formation. Saline water also boils at a higher temperature than fresh water.

Ions dissolved in water have numerous ramifications. For example, when no ions are added to water it is an effective electrical insulator (electrical conductivity of 5.5×10^{-10} S/m compared to something like copper with 6.0×10^7 S/m) since water molecules themselves cannot transport charge. But even ultrapure, "deionized" water is not truly ion-free. Startlingly, liquid water will spontaneously auto-ionize by

Table 1.1.	Saline water classifications.
Fresh water	<1,000 ppm
Slightly saline water	1,000–3,000 ppm
Moderately saline water	3,000–10,000 ppm
Highly saline water	10,000–35,000 ppm
Seawater	~35,000 ppm
Brine	>40,000 ppm

having two water molecules react to form one hydroxide ion (OH^-) and one hydronium ion (H_3O^+). The hydroxide ion is the primary constituent of bases while the hydronium ion is the primary constituent of acids. This process is driven by thermal energy in the water, so the absolute concentration of these species will be temperature-dependent, although the relative concentrations of hydroxide and hydronium from auto-ionization in pure water will always be equal. The concentration of hydrogen ions (H^+) (or, rigorously, the thermodynamic "activity" of hydrogen ions, which is an effective concentration that also depends on temperature, pressure, and other factors), which generally exist as hydronium ions in water, defines the pH of the solution. Specifically, pH is the negative logarithm of the hydrogen ion concentration (or activity). Being logarithmic, a change in pH of 1 represents a 10-fold difference in H^+ ion concentration. Pure water has a neutral pH of 7, with acids having lower pH (more H_3O^+) and bases having higher pH (more OH^-). Shifting away from neutral pH means that there are more ions, and therefore more electrical conductivity since the ions can carry charge. Similarly, if there are other ions added to the water, such as from salts, the conductivity will also rise. Electrical conductivity, therefore, is an indirect measure of the overall concentration of ions in the water. Rainfall generally has a pH around 5.6 — slightly acidic as a result of dissolved carbon dioxide and nitrogen/sulfur oxides from the atmosphere. Seawater, in contrast, has a slightly basic pH around 8 thanks to bicarbonate ions (HCO_3^-). (Ocean pH has been steadily decreasing for decades as a result of absorption of increasing amounts of carbon dioxide from the atmosphere originating from burning of fossil fuels, which creates carbonic acid when mixed with water.) pH is among the most important metrics of water quality. When the pH changes in municipal water supplies, it can have dire consequences. For example, when Flint, MI switched their water supply off the Detroit system in favor of water from the Flint River in 2014, the corrosive lower pH destroyed protective coatings within supply pipes and leached dangerous amounts of toxic lead into the water.

Water is essential to all life on Earth. It is a simple molecule, containing just three atoms, but these humble little fellas are beguiling. As you can see, despite the fact that water is seemingly everywhere and we interact with it in so many ways every day of our lives, it is far from a simple or boring material. On the contrary, it is astoundingly complex and multifaceted. Water, especially in liquid form, stubbornly holds onto some of its secrets, confounding scientists and engineers even today.

Water volume and flow

Stationary water is relevant for all sorts of reasons, but it is the movement of water from one place to another that enables most of its use, whether it be for irrigation, power generation, or residential use. Before delving into a discussion of the movement of water, let's first go over common units used to quantify volumes and flow of water. Table 1.2 lists equivalent values for these parameters in various common units to assist in conversion.

When discussing water use in a home, units such as gallons or cubic meters are convenient. On an agricultural scale, acre-feet is more common. An acre-foot is the volume of water that would cover one acre of land one foot deep, which is a lot of water indeed. In municipal systems, million gallons per day or million tonnes per day (equivalent to million cubic meters) are most common.

Fluid mechanics is the field of study encompassing the flow of water, and here we will review some of the basic principles of fluid mechanics as

Table 1.2.   Common units for water volume and flow.

Volume		Flow	
Gallon (g)	325,851	Gallons per minute (gpm)	694.444
Liter (l)	1,233,480	Million gallons per day (mgd)	1
Acre-inch	12	Liters per minute (lpm)	2,628.76
Acre-foot	1	Million liters per day (mld)	3.78541
Cubic meter (m^3)	1,233.48	Cubic meters per second (cms)	0.052616
Cubic foot (ft^3)	43,560.0	Cubic feet per second (cfs)	1.85814

they relate to water since these principles — coupled with the properties of water we outlined earlier in this chapter — are the foundation of essentially all water technologies. Perhaps the most basic principle is the continuum assumption. Liquid water is of course comprised of many discrete molecules that can move largely independently from one another. Continuity, however, treats the water as a continuous material. Properties such as temperature, density, pressure, and velocity are presumed to be well-defined at all points and to vary continuously from one point to another. Another assumption generally applied to water is that it is a "Newtonian" fluid. This means that regardless of the forces acting on water, its viscosity remains the same. Stirring a non-Newtonian fluid can cause it to be less (or more) viscous, as with ketchup and paint (or cream and quicksand). No matter how much you stir water, it just keeps flowing.

The most general mathematical treatment for describing the motion of fluids such as water is the Navier–Stokes equations (named in honor of brilliant physicists Claude-Louis Navier and Sir George Stokes). These differential equations describe the balance of forces acting at any given region of a fluid. For most real-world situations, water flow is sufficiently complex that these equations can only be solved with the help of computers, a branch of science known as computational fluid dynamics. For our purposes, we will focus on simplified concepts derived from the Navier–Stokes equations that can be employed safely in most applications with water. Generally, flow can occur in two regimes: laminar or turbulent. Laminar flow, which tends to occur at lower velocities, describes a smooth, orderly condition in which water flows in parallel layers with no lateral mixing or cross-currents. Increasing the velocity beyond the so-called critical velocity initiates more chaotic turbulent flow, characterized by eddies or swirls of water with substantial mixing.

Two specific cases that come up time and again with water are (1) flow through pipes and (2) flow through porous materials. In 1738, Swiss mathematician and physicist Daniel Bernoulli published *Hydrodynamica*, a book in which he described what is known today as the Bernoulli principle: that an increase in the speed of a fluid occurs simultaneously with a decrease in pressure. His principle is the foundation underlying equations that describe the flow of water through pipes. Things become more complicated when the pipes change diameter or bend, but effective formalisms exist to treat these common situations. Another factor that cannot be ignored is friction. There are various origins of friction as water flows through a pipe, most notably viscosity and turbulence of the water and interactions with the pipe walls. This means that pipe diameter and wall roughness affect the amount of friction experienced by the water. Electrical energy is dissipated when an electrical current flows through a resistor; similarly, energy is dissipated when water flows through a pipe. In the electrical circuit this is manifest by the potential drop across the resistor; in the case of the pipe, the flow causes a pressure drop along the pipe.

In a sense, one can think about water flow though porous materials as water flowing through a really messy network of tiny pipes. Flow through porous materials was first studied in detail by a 19th century French engineer named Henry Darcy. He built a pressurized water distribution system that carried water brought through a covered aqueduct from Rosoir Spring to the city of Dijon, France. The system was fully closed and driven by gravity alone (no pumps) with sand acting as a filter. After retiring from municipal work, he carried on his interest in water and sand. He performed experiments in which he observed that the flow rate of water passing through a plug of sand was directly proportional to the pressure across the sand, with the proportionality scaled by the "permeability" of the sand and the water's viscosity. Today this relation is known as Darcy's law (it can be applied to many fluids and porous media), and the standard unit of permeability is even named in his honor. Permeability is quite an important aspect of the movement and purification of water in both natural and artificial systems. It is primarily the structure of pores within a material that determines its permeability, specifically, the "porosity" and "tortuosity" of those pores. Porosity is simply the fraction of the material's volume that is comprised by empty space, or pores. Tortuosity is defined as the ratio of the average length of the convoluted path that the fluid must travel to traverse the material to the material's thickness. If water can take a nearly straight shot through a material like a car on an open highway, the tortuosity is low, and if it has to meander this way and that the whole time like a car navigating crowded surface streets, the tortuosity is high.

Now that we have a basic understanding of the nature of water, its rather unusual properties, and how it moves, we can look at water's essential role in the living world.

02 Water is Biology

What are two bodily functions that you take care of when you wake up in the morning? Lots of folks will say that they use the bathroom and get a drink of water. Your body is using water as a fluid to manage biological functions. In the case of the bathroom, you are expelling degraded nutrients from food and physiological activities that are potentially detrimental or are no longer needed. In the case of drinking water, you are replenishing the supply so that you can continue biological activity. There are four basic requirements for humans and animals to survive: oxygen from air, water, food, and shelter from the elements. In terms of survival, we can only survive a few minutes without oxygen, depending on the ambient temperature, only a few days without water, but potentially weeks without food. Our need for shelter is based on exposure to temperature extremes, and there are societies that exist for millennia without shelter as defined in our culture.

Maybe excepting jellyfish, we think of living species as solids. If you analyze the content of an organism, you quickly learn that they are actually about 65% water. Yes, 65% water, similar to the amount of the earth's surface area covered by water. Our brains are almost three-quarters water. In fact, biological cells could be considered bags of water with some other "stuff" either dissolved or suspended in the water. (*Star Trek: The Next Generation* fans may recall the Season 1 episode in which a living crystal the *Enterprise* encounters during its voyage derisively refers to

Captain Picard — and all humans — as "ugly bags of mostly water.") Cells can have a water content as high as 90%. With eukaryote cells, such as humans, animals, plants, and even yeast, the cell has subcellular components present in the cell body. The subcellular compartments include the nucleus, mitochondria, ribosome, chloroplasts in plants, and others. They, too, appear to be bags of water, but may have a lower overall water content. The more ancient forms of cells including prokaryotes are also simply bags of water without subcellular compartments.

From the molecule to the cell

So why is there so much water in living things? Water plays many useful roles in biology.

In an organism, some cells are mobile such as blood cells, while others are linked together in a structure such as organs, bones, muscles, and nerves. It's pretty important that certain cells be able to get around — like when you cut your finger chopping a cucumber, it's what enables the skin to heal itself. Water is the basic fluid medium for transporting the mobile cells around our bodies. Water is also the primary component in the interstitial space between cells.

It is the fluid carrier to deliver nutrients, oxygen, and energy. Water transports wastes and toxins in cells and whole organisms. Water carries hormones and other signals from one part of the body to communicate with another part. Water provides the medium for the many reactions required in life. These reactions include energy production such as generating ATP (adenosine triphosphate) from glucose (sugar), synthesis of biological polymers such as deoxyribonucleic acid (DNA), ribonucleic acid (RNA), proteins, starch, and cellulose. Just imagine all those innumerable kinds of molecules diffusing around, reacting and decomposing — all dissolved in one miraculous solvent: water.

It is important to note here that water does more than just sit on the sideline and let all the other molecules participate in the chemical reaction game. Water, too, participates in reactions in biology. It is a reactant in some cases (involving "hydrolysis"), and a product of others (involving "condensation"). Digestion, photosynthesis, and aerobic respiration are just a few of the reactions in living things in which water plays a central role. In photosynthesis, for example, carbon dioxide and water are transformed into glucose and oxygen with help from light energy. Aerobic respiration is the reverse process, which releases energy.

In the previous chapter, the unique chemical and physical properties of water were described. The ability of water to form hydrogen bonds, in particular, was highlighted. Hydrogen bonding in water provides the backdrop for one of biology's most exciting features. Hydrogen bonding in water allows biological molecule to self-organize into complex structures and function as microscopic machines. Biology's building blocks include amino acids for proteins, nucleic acids for DNA and RNA, sugars for polysaccharides, and fatty acids and lipids for lipid-bilayer membranes. In each case, the unique chemical functionality of water enables biology to create complex structures and provide highly specific and selective functions that allow life to function as we know it.

Perhaps the most famous example is DNA, which is comprised of a bunch of nucleic acids. DNA forms a double helix in which two strands of nucleic acid chains wrap around each other. DNA stores our genetic information — essentially the rulebook for life. A DNA double helix is a very stable structure and literally lasts a lifetime. (In fact, under proper conditions, it will survive for millions of years.) If you ask most folks who discovered DNA, they will likely point to American biologist James Watson and English physicist Francis Crick. Not so. In fact, DNA was first identified back in the 1800s by a Swiss chemist named Friedrich Miescher. After this discovery, other scientists performed experiments that revealed additional details about the DNA molecule, such as revealing its chemical constituents and how they are bound together. Watson and Crick built on this foundation with their famous discovery that DNA is arranged in a double helix, a breakthrough that relied heavily on x-ray diffraction data collected by Rosalind Franklin and Maurice Wilkins. In Watson and Crick's 1953 publication in *Nature*, they stated in sardonic fashion "It has not escaped our notice that the specific pairing we have postulated immediately suggests a possible copying mechanism for the genetic material". Individual sections or genes can be copied from DNA as the templates for all of the proteins of life. The entire DNA strand can be duplicated to provide the genetic information for a new cell or a new organism. To do these remarkable functions, DNA is comprised of three sections. DNA has a nucleic base attached to a deoxyribose sugar attached to a phosphate linkage (Figure 2.1). The phosphate linker attaches to the next base to form the strand. (Human DNA is about three billion base pairs long. It is tightly bundled into the nucleus in the cell, but if stretched out would span about a meter in length.) The base is hydrophobic (water-fearing) and avoids water. The sugar and phosphate are hydrophilic, or water-loving. In the double helix, the ribose and phosphate are exposed to water surrounding the DNA while the bases from opposite strands are brought into close proximity on the interior of the molecule to avoid exposure to water. The bases, called purines and pyrimidines, are linked through either two or three hydrogen bonds. The hydrogen bonds are at the essence of the fidelity of the genetic information. The relationship

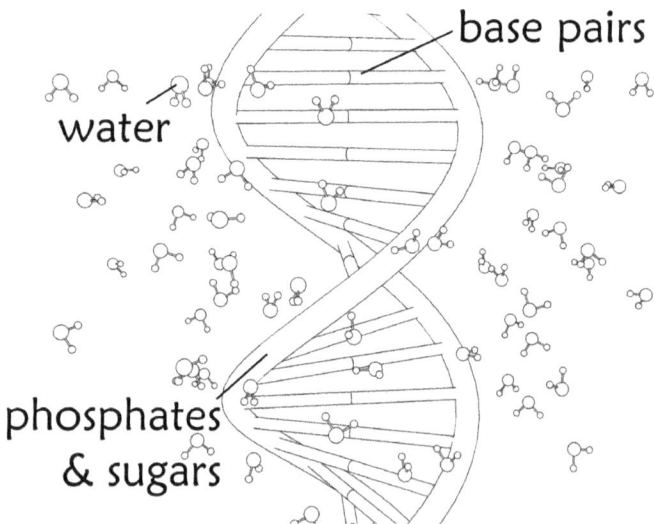

Figure 2.1.   Structure of DNA is dictated largely by interactions with water.

with water between the hydrophilic shell and hydrophobic core makes all this possible.

The relationship between proteins and water is even more intimate than the case of nucleic acids. Proteins are the primary workers in biology, driving chemical reactions, transporting nutrients, sensing signals, and establishing structure. Protein chemistry and protein biology are among the richest subjects in all of science. The thousands of functions that proteins perform are driven by the *billions* of structural features of proteins. These structures, in turn, derive from the interactions of protein chains with water. There are twenty common amino acids that serve as building blocks for virtually all proteins in living species, whether prokaryotes (e.g. bacteria), fungi, plants, or animals. Some amino acids are hydrophilic and some are hydrophobic, and others are somewhere in between. There are amino acids that are positively charged and that act like bases, and there are amino acids that are negatively charged and that act like acids. These chemical properties are induced by water and enable "primary" amino acid chains (i.e. proteins) to form several types of "secondary" structures. The two major secondary structures are alpha helices and beta sheets (Figure 2.2). Hydrophilic amino acids are more likely to fold into alpha helices, which are single-stranded, distinct from the more rigid DNA double-stranded helix. In the alpha helix, the hydrophilic amino acids are exposed to the surrounding water. Although perhaps not as famous as Watson and Crick, two years earlier, Linus Pauling predicted

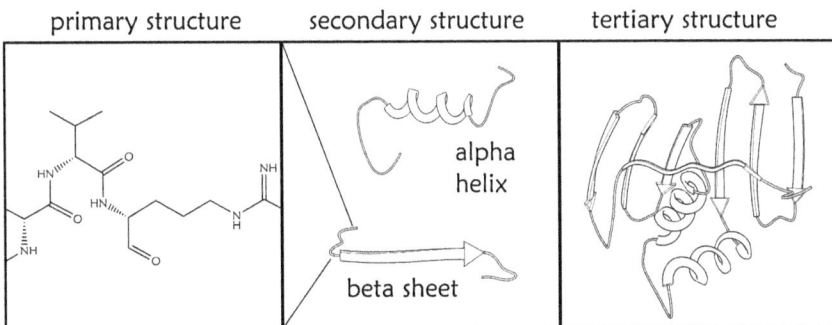

primary structure   secondary structure   tertiary structure

alpha helix

beta sheet

Figure 2.2.   Hierarchical protein structures.

the structure of the protein alpha helix, using his exquisite knowledge of hydrogen bonding and without any x-ray diffraction data to inform his insight. In a beta sheet, the hydrophobic portions of the protein are shielded from the water because they are located near the interior of the structure. By themselves these two secondary structures don't create the rich diversity of structures and functions exhibited by proteins. Rather, proteins spontaneously assemble into some of the most complex structures in nature. Through bridging strands, the secondary structures fold into rich and diverse "tertiary" structures. These tertiary structures are complex three-dimensional structures that form pockets, channels, reaction sites, binding sites, and virtually any other possible geometry for biological function. Interactions between intra-protein tertiary structures can even form inter-protein "quaternary" structures. Tertiary (and quaternary) structures in proteins form pockets that are designed to only interact with very specific species such as amino acids, nucleic acid sequences, sugars, nutrients, signals, or even other proteins. The twenty basic amino acids can form almost an infinite number of tertiary structures, each with its own specificity and function. And none of this would happen without water. Ultimately, life has used the physiochemical properties of water to create a rich and diverse environment for carrying out any potential function or reaction it may need.

Another critical structure in biology derived from interactions with water molecules is the lipid bilayer (or phospholipid bilayer). This structure is an extremely thin membrane made of a mere two layers of molecules. Lipid bilayer membranes have a thickness of about 2–4 nanometers (billionths of a meter) and encapsulate human cells as large as 100,000 nanometers. Lipids are molecules that have an end-group made of a hydrophilic, sugar-like structure called glycerin. The other end of the molecule has three long-chain fatty acids that are oil-like and repel water (hydrophobic). In the presence of water, lipids form a sheet with the glycerin parts facing the water, and the long chains facing inward to avoid contact with water. This behavior lets lipids form a barrier that can contain and control the contents both inside the membrane and in the space around it. The cell membranes of nearly every living organism are made of these lipid bilayers, as are the barriers surrounding the nuclei and other structures within the cells. Despite being phenomenally thin, lipid bilayers are well suited to keeping ions and molecules where they should be. These membranes are impermeable to most molecules that dissolve in water, which is, as we've discussed, a lot of molecules. So how does stuff get into or out of a cell? Membranes have embedded channels to control both intake and release of nutrients, wastes, signals, or any product of the cell. These channels are assembled from proteins. The channels are frequently hydrophilic, bathed in water. Therefore, the hydrophobic membrane has components that allow water soluble species to travel in and out of the cell (Figure 2.3). The channel proteins are essential for a cell's survival and also for its contribution to the overall organism.

Another type of protein that sits within a lipid bilayer is an ion pump, which the cell uses to regulate salt concentrations and pH by transporting ions across the membrane. Yet another class of important channel proteins are known as aquaporins. These fascinating fellows are miraculously good at transporting water through the membrane. Most cells move water in and out through good, old fashioned osmosis (see Chapter 8). Some cells,

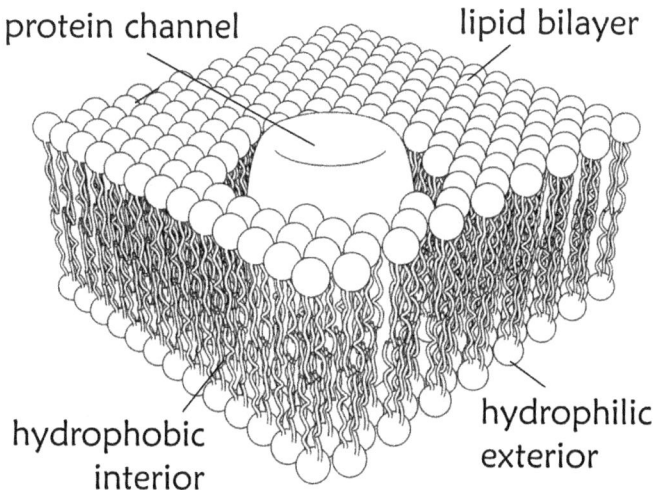

Figure 2.3.   Lipid bilayer membrane with protein channel.

though, need a lot more movement of water than slow osmosis through lipid bilayers can provide, and this can be tricky if you want lots of water but not other stuff, like salt. Plants that soak up water through their roots are an obvious example. Our kidneys need this sort of function too. Enter aquaporins. Scientists are still working out the details of exactly how these proteins are so efficient and selective for water transport — and some are even trying to use them to create artificial membranes for water treatment.

Like with DNA, if it weren't for water, lipid bilayers wouldn't form, and we wouldn't be here. But water's importance in biology does not stop at the cellular level. In the next section, we will explore some ways in which water interacts with larger biological systems.

From the cell to the organism

The biopolymers we've discussed above store and transmit information, do work, and provide structure and barriers; thanks to water, they provide critical functions for cells. In complex, multicellular organisms, different types of cells organize into tissues and organs. Tissues such as muscles are comprised of just a few cell types and play rudimentary, but frequently macroscopic, roles in an organism. Consider muscles. They respond to an influx of salt ions, which change the conductivity of the water in the cell. The change in conductivity triggers an electrical response causing long muscle cells to contract into more compact spheres. The coordinated change in the shape of a set of microscopic muscle cells induces macroscopic movement of a portion of the organism, such as bending your elbow. There is water again, playing a central role. Now consider an organ, say the heart. Organs play highly specific roles within an organism. The heart is almost entirely a muscle with some neurons for receiving signals on what to do. Coordinated contraction and expansion of heart muscle tissues causes a series of chambers to suck in oxygen-depleted blood from veins, send it to the lungs for oxygenation, and push the oxygen-rich blood into arteries. While the heart is about 73% water, it functions as a soft solid and is able to compartmentalize blood, and direct it to and from blood vessels. Blood, by the way, is greater than 90% water.

In Chapter 1 we described how hydrogen bonding of water leads to a high heat capacity, boiling point, and melting point. Life takes advantage of these properties, using water as a temperature buffer. Because it isn't easy to change the temperature of water, having a bunch of water around keeps the temperature relatively stable — sort of like a coffee thermos, except wetter. All of the chemical reactions that take place in living things have rates that are intimately sensitive to the surrounding temperature, so a stable temperature keeps all those processes from running amok.

There is another aspect in which water is central to biology, and it's so obvious you could almost forget: lots of things live in the water.

Biology from freshwater to salt water

While humans — and your dog or cat — depend on freshwater, more than four billion years ago, life started out in salt water. Still today, more species live in salt water than freshwater, and as we see in fish such as salmon, some spend parts of their lifecycle in both. The interactions between biology, life, salt, and water are nuanced. While we think of humans as freshwater beings, we are actually somewhat salty. Our physiological salt content is almost 1%. As discussed in Chapter 1, seawater is about 3.5% salt. So while we are not quite like seawater, we are much saltier than drinking water (0.3% salt). When you consider the spectrum from glacial runoff through freshwater rivers to ground water, brackish water, seawater, and isolated briny bodies such as the Dead Sea, water on Earth exists over a wide range of salinities. Amazingly, life has thrived in water across this entire range of salinity.

How does life adapt to the available salinity? This ties in to the balancing of osmotic pressure discussed in Chapter 8. Salt wants to surround itself with water. When the salt concentration is higher on one side of a semi-permeable membrane, water flows from the less salty side into the salty side to dilute the salt. We describe this phenomenon as osmotic pressure. Organisms strive to balance their internal salt content with their surroundings. So a fish surviving in the ocean has a much higher salt content in its tissues than a freshwater fish. Humans cannot survive by drinking seawater. We drink water at much lower salt contents than our bodies. We balance the salt content with the food we eat. How do fish such as salmon live in both fresh and salt water? They use a process called osmoregulation. When salt content is too high for survival, they collect excess salt in their kidneys and excrete it through their gills. When salt content is too low to survive, they excrete excess water through their kidneys. In other words, it requires a lot of energy to balance salt and water. While salmon have been bred to spend their entire life in freshwater, they are most suited to mature and grow in saltwater, and only spawn and die in freshwater. If the water is super salty, at some point organisms like fish or plants can no longer overcome the osmotic pressure, and they die. The Dead Sea is, as the name would suggest, such a place. That said, the Dead Sea is not totally dead. Life finds a way. Several species of bacteria, microscopic algae, and fungi that have adapted to hyper-salinity conditions live there. Deep sea brine pools also can teem with life. Bacteria feed on methane seeping from the sea floor, and the bacteria, in turn, provide food for creatures such as mussels. On the other end of the salt spectrum, if humans or other organisms consumed only distilled water with a very low salt content, our health would suffer, even potentially leading to death. The salinity of our tissues would decline (osmosis again) and water balances would get out of whack. In the next chapter, we will explore water's role more broadly in the natural world.

03 Water is Nature

The foundation of water in our consciousness lies in nature, the environment, and ecosystems. While access to drinking water frightens many and flooding unsettles others, we imagine water as lakes, rivers, streams, and the ocean. We find solace when we are close to natural water bodies. We seek proximity to water for contemplation, in our free time, and during vacations. This behavior is universal across societies and represents a core desire in humans. The authors spent much of their adult lives a short distance from Lake Michigan. Although the water is only warm enough to swim comfortably for a few weeks a year, the lakefront attracts people yearround. What is it about water in nature that attracts us?

If we think back to Chapter 1, water has unique physical and chemical properties that make it behave distinctly from other substances. There are many aspects of water, from floating ice to the ability to dissolve many substances and repel others that empower water to provide a rich and varied template in nature. If we think back to Chapter 2, water enables the molecules of life to take on complex structures and functions. Water is necessary for all life. The varied physical conditions of water, from salt concentration to the ability to retain heat and buffer temperature variations, enables living species to create unique niches in water. Each body of water supports a unique ensemble or ecosystem of living species. These environmental niches are based on water's properties and enable life to form relatively stable communities. Rich and varied communities provide sustenance for individual species, provide a means to share and balance resources, recycle waste, and create a sustainable environment.

The cycles, locations, and ages of water

As we described in Chapter 1, the water molecules with which we interact largely date to the formation of Earth, billions of years ago. In the context of nature, though, one can look at water's age from a different perspective. The real questions are: where is the water, how and why did it get there, and how long does it persist in those places? Taken together, these questions harken back to something we all learned about in grade school — the water cycle. As it is a cycle, we can start our discussion at any arbitrary point in the water cycle, and we will eventually return to the same point. Let's consider volumes and start at the largest source.

Our planet has a breathtakingly large amount of water. Figure 3.1 shows how this water is distributed on Earth. Nearly all of it — approximately 97% — is located in the oceans. Thus, salty, or saline, seawater is the ultimate starting and endpoint for water. After seawater, frozen water in the form of ice caps, glaciers, and long-lasting snow represent a little less than 2% of all water. Interestingly, frozen water

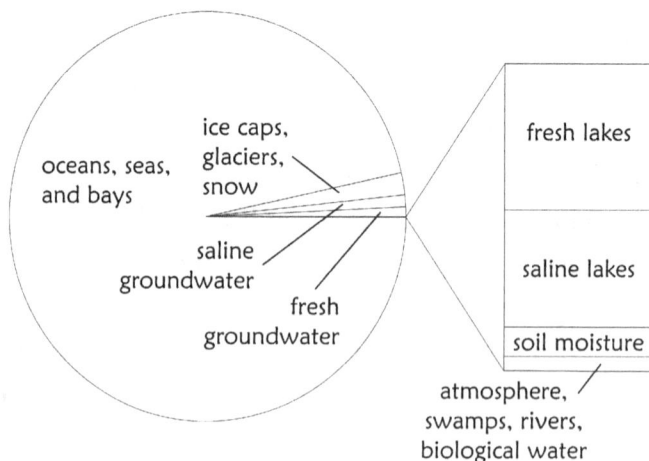

Figure 3.1.   The relative volumes of major water sources on Earth.

accounts for nearly 70% of all freshwater on the planet. After that, about 1.5% of water is sequestered in aquifers or "groundwater" across a range of salinities. Fresh groundwater represents the lion's share of all liquid freshwater. Every one of the rest of the water sources we think of, from lakes to rivers to soil moisture to air moisture, each account for less than 0.01% of total water and less than 0.05% of freshwater present on Earth. Think about that for a moment. Lakes and rivers seem so massive, but they represent an almost imperceptibly tiny fraction of the planet's water. The largest accessible freshwater source on the surface of the earth is the Great Lakes, which account for 22% of the world's surface freshwater and 84% of North America's freshwater. Lake Baikal, located in southern Siberia, contains a bit more surface freshwater than the Great Lakes, and is home to a rich diversity of aquatic life, many unique to the area. This region of the world is much less populated than the Great Lakes region, and it is not feasible for major water users to access this resource.

From this perspective, freshwater as a resource starts to look both more precarious and precious. If all of the water on Earth was collected in a sphere, it would have a diameter of under 1,000 miles. Even though the oceans cover 70% of the planet's surface, their average depth of two miles makes them a thin coating in terms of the total volume of Earth. If we consider only the surface freshwater, it could all be encapsulated in a sphere of about 30 miles diameter, basically a bubble over any large city. That's all there is.

Starting from the ocean as either the endpoint or starting point, let us consider water cycles and residence times. Residence time is a measure of how long water stays in that particular stage of the water cycle. Just a small fraction of the oceans' volume evaporates from the surface, so typical residence time there is thousands of years. First and foremost, every aspect of the water cycle is driven by energy. Ultimately, that energy comes from only two sources — sunlight and Earth's interior. Sunlight drives all the routine functions, and interior energy causes some isolated events. When sunlight strikes the surface of the ocean, some is absorbed. Photosynthetic organisms add tints to the water near the surface, increasing sunlight absorption. Some water evaporates into the atmosphere. Wind sweeps more water-saturated air from the ocean surface, letting drier air come in contact with the ocean surface to further encourage evaporation. Many factors control evaporation rates, including the temperature of the water, temperature of the air, the speed that saturated air is transported away from the surface, sunlight, and water salinity. (In Chapter 8, we will address how climate change is altering this process.) In Chapter 1, we learned that evaporation requires energy and its occurrence cools the surroundings. Therefore, evaporation acts as a buffer slowing down its own rate. Evaporation also occurs from surface freshwater bodies. While freshwater evaporates more readily than saline water, the surface area of lakes and rivers is dramatically smaller than the oceans'.

Plants and trees also send water to the atmosphere through evapotranspiration. Water transports nutrients from the soil through the roots to the leaves, and most of that water is lost to the atmosphere. The atmosphere is agnostic to how water arrives there. No matter how the water gets into the atmosphere, it eventually collects in clouds. Clouds are formed by mixtures of dust particles and condensed water vapor. Cooling decreases the water solubility in the atmosphere, and eventually condensed water precipitates from the sky as rain, snow, sleet, hail, or fog (or graupel...yeah, that's a real thing...look it up). Just as evaporation requires heat and cools the surrounding environment, condensation releases energy and warms the surrounding environment. We all experience this process firsthand. We sweat to provide water for evaporation to cool our skin. In humid environments, the sweat pools and makes us feel warmer. We also experience this phenomenon when we enter a warm home from the cold outside. Our cold glasses condense water and fog up. Based on the balance of evaporation rate, water concentration, and precipitation rate, the residence time in the atmosphere is typically only days to weeks. This is one reason why precipitation rates are higher near water bodies — think of rainy Seattle or the snowbelt east of the Great Lakes. Just based on the relative fraction of the Earth's surface area, most precipitation falls on the oceans, looping the cycle back to where we began. Some, however, falls on the land or in freshwater bodies, triggering the other aspects of the water cycle.

Remember that a primary force acting on water is gravity. Water flows downhill. No matter where it falls, liquid water makes a pathway toward the ocean. If it falls in the mountains, small streams aggregate water into rivers and lakes as we describe later in this chapter. Some of the water will flow all the way to the ocean whereas some will pass through porous sections of rock and soil and infiltrate into the ground. Watch a river flow rate, estimate the distance from the head of that river to the ocean, and you could determine the approximate residence time of the water. The age of water in lakes is dependent on the depth and outflow from the lake. Described later in this chapter, the age can span orders of magnitude. As with the ocean, underground aquifers could be considered an endpoint with some residence times reaching hundreds of thousands of years. There is a sidetrack for water that precipitates on land. Near the poles and at high altitudes, the snow or frozen rain can deposit on the land and stay for very long periods of time. For examples, the snow is sections of Antarctica has packed down, season after season, until it is hundreds of feet thick, and the lower parts may have been there for millennia. This water returns to the ocean only when disrupted, such as the calving of an ice shelf. Among the most critical risks from climate change is melting of the snow and land-based ice near the poles and in glaciers. Melting water will make its way to the ocean and cause relentless sea level rise. (Check out Chapter 8 for more on this topic.) The basic components of the water cycle are shown in Figure 3.2.

Water ecosystems and nature

Let us ask: What is nature? One can describe nature as the interplay between living communities and the inanimate structures that support life. A common theme is that "nature" represents systems with limited disruption or intervention from humans. If we base the value of nature on its ability to sustainably support life, then water is absolutely necessary

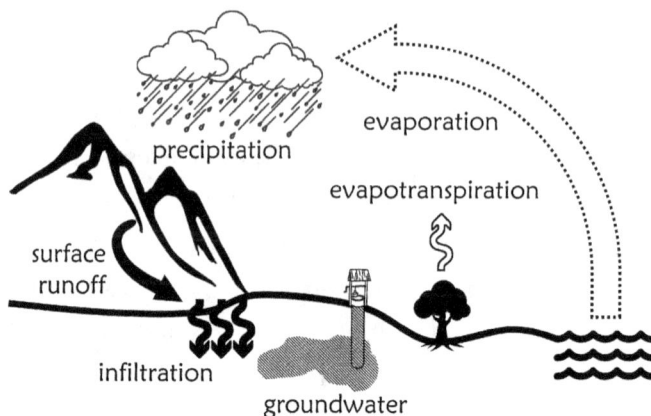

Figure 3.2.   The water cycle, from the ocean to the atmosphere to the land and back to the ocean.

for nature and water *is* nature. We consider four representative natural water systems: the ocean, a river, a lake, and a wetland to outline some of the diversity of nature.

Life started in the ocean. As we mentioned above, the oceans cover more than 70% of the surface and account for 97% of all of Earth's water. Traditionally we think about "the seven seas." We consider the Atlantic Ocean, Pacific Ocean, Indian Ocean, and Arctic Ocean as separate bodies. In fact, they are a continuous water body. Buckminster Fuller, the architect of the geodesic dome, designed a dymaxion projection map of the world to represent Earth as either one island (projected around the North Pole) or one ocean (projected around the South Pole) that surrounds Antarctica. ("Dymaxion" is a portmanteau of the words dynamic, maximum, and tension.) The ocean's salinity is about 35,000 parts per million, or 3.5%, and except near melting glaciers or river basins is generally consistent across the world. While the salt in the ocean prevents it from freezing, the surface temperature varies widely from less than 30°F in the winter Arctic Ocean to greater than 90°F in the tropics. The temperature decreases rapidly with depth, falling by 10°F below ~300 feet and further declining to less than 30°F below 3000 feet, worldwide. With the average depth of the ocean's floor being greater than 10,000 feet, most of the ocean is cold and dark. Little to no sunlight penetrates beyond 600 feet. While the great depth makes it difficult for visible light to pass all the way to the ocean floor, our other sensor and detection tools also suffer. Radar, sonar, microwaves, radio waves, and other probes cannot pass by so many densely-packed water molecules. Therefore, most of the ocean floor is unexplored, un-probed, and unknown. Think of the many mapping resources available, satellite imagery, and GPS. We have detailed knowledge of the land surface at a sub-meter scale. Away from the continental shelf, our knowledge of the ocean floor is really only known at the kilometer scale, orders of magnitude less than on land.

Most of our experience with the ocean's natural systems are in coastal regions along the shallow continental shelf. Living communities or ecosystems near the coast are varied and rich. Minerals and nutrients are carried into the ocean from the land by water in rivers or by air with the wind. While the oceans support life in both cold and warm regions, the variety and density are richer in warmer coastal waters. These communities create a food supply chain that at its base is supported by photosynthetic organisms including plankton and algae. Photosynthetic organisms depend on the minerals and nutrients that are provided by the land. The photosynthetic organisms provide sustenance to small and large species, including a range of shellfish, grazing fish, and even some whales. The small creatures provide sustenance to more muscular predator fish. Their survival is dependent on the grazing species and ultimately on the photosynthetic organisms. In warm regions, the ocean supports coral ecosystems, a rich community of stationary and mobile fauna and fish.

Away from the coasts, the concentration of nutrients and minerals transported by rivers and the wind declines rapidly. Mid-ocean water, miles from the coast, is relatively devoid of life. While the oceans are a continuous system, natural communities concentrate in coastal waters and are physically separated by the mid-ocean. This arrangement has created isolated communities that have distinct populations in various coastal ecosystems, even if the temperature and environmental conditions are similar. These communities developed and evolved in isolation from each other. Human intervention has disrupted this isolation. Sometimes by accident, boats carry and transfer fauna, attached shellfish, crustaceans, or fish in their ballast water or on their hulls between ecosystems. Ecosystems that existed in isolation now are populated with common species. This disruption can be severe if the introduced species has no natural predator, and the invasive population explodes. Also, if the foreign species introduces disease vectors or is a predator in the adopted community, it can cause the collapse of the established ecosystem. At times, the introduction of foreign species has been intentional. One driver for this behavior is to add a desired food source to an existing ecosystem to supplement populations that humans consume or that contributes economic benefit. While foreign introduction may provide short-term benefits, sometimes there are unintended consequences. Even a planned introduction of a non-native species has the potential to disrupt virtually any natural ecosystem. The Great Lakes have experienced a series of disruptive introductions, including sea lampreys, round gobies, alewives, salmon, and zebra mussels. The sea lampreys and alewives slowly migrated up the shipping channels built to promote Midwest trade. Other species arrived in the ballast water, mostly from Europe and the Caucus region. Pacific salmon were introduced to support the gamefish industry that collapsed from the decimation of lake trout by the sea lamprey invasion. There is great concern that Asian carp will escape the Mississippi River/ Illinois River system and enter Lake Michigan through the Illinois Sanitary and Shipping Canal. If that happens, it will be just the latest population disruption among a series dating back nearly two centuries.

Oceans are an essential component in the world's water and natural cycles beyond serving as a haven for life. They are the endpoint for transport of virtually all water in the world. A small fraction of water ends up sitting for a long time in landlocked water bodies, which create highly saline lakes and seas including the Dead Sea in the Middle East, the Great Salt Lake in Utah, and the Salton Sea in California. There are landlocked saltwater bodies in every continent on Earth, including Antarctica. The most saline water bodies in the world are the Gaet'ale

Pond located in the Afar Depression in Ethiopia and Don Juan Pond in Antarctica. These hypersaline bodies are both over 40% salt, more than ten times saltier than the ocean! As the endpoint, nature's water, minerals, nutrients, wastes, plastics, and all sorts of other stuff wind up in the ocean. One might wonder, if all of this water heads to the ocean, why doesn't the ocean continue to rise? Evaporation, of course. But if so much water evaporates, why doesn't the ocean get as salty as hypersaline lakes? Many of the minerals are either incorporated into shells or, because of limited solubility, slowly precipitate to the ocean floor. Thus, we see a natural transfer of minerals from the mountains to the sea floor. Over millions of years, tectonic plate activity raised the ocean floor to form new mountains. The action of water and gravity ultimately return the minerals to the ocean floor in a natural cycle of water and minerals.

The oceans are not stagnant bodies. There is constant motion across multiple distance scales. When you approach the seashore, the first thing you notice are waves. Essentially, these oscillations are vertical motion of water — the water moves up and down, not forward or backward. Waves are really energy movement in response to wind and interactions with the seafloor. Sit by the seashore for an hour or two, and you observe tides, a much larger level of motion in the ocean. Tides result from the gravitational pull of the moon and, to a lesser extent, the Sun on Earth. High tide occurs twice a day, when the moon is directly above or below. Low tides occur when the moon is at the horizon. Tides are more pronounced near the poles and less pronounced near the equator. Basically, the moon exerts a gravitational pull on Earth, and the fluid water responds. Considering that the diameter of Earth is about 8,000 miles, a tidal rise of 5–10 feet represents only a tiny distortion in the shape of Earth, 0.2 parts in 10 million.

At an even larger scale are ocean currents. Ocean currents result from a range of forces, including the rotation of Earth, prevailing wind patterns, and interactions with jagged coastlines. Ocean currents, such as the Atlantic Ocean's Gulf Stream, circulate in a clockwise motion north of the equator and in a counterclockwise direction south of the equator. The

Gulf Stream carries warm water from the Caribbean Sea up the Atlantic coast of the U.S. and Canada and eventually to the British Isles. The Gulf Stream provides warm water to Britain, resulting in a more temperate climate than other countries at a similar latitude. Similarly, the California Current transports cold water from Alaska to the California coast, making the oceanfront from San Francisco to San Diego much colder than the East Coast. Beyond ocean currents, there are major but sporadic water events across the oceans. Tropical storms are called typhoons in the Pacific, hurricanes in the Atlantic, and cyclones in the Indian Ocean and South Pacific. They develop when warm air over the ocean creates an extreme low-pressure zone. The energy is derived from the evaporation of seawater by the warm air. Moist, warm air circulates in a counterclockwise motion and picks up moisture and speed while it moves across the ocean. With wind speeds reaching 150 miles per hour and widths of hundreds of miles, major tropical storms are among the most destructive events in nature. With improved tracking and modern weather modeling, we are learning to predict landfall and pathways for these monsters. Models of climate change strongly indicate that warming ocean water will cause the world to experience more severe tropical storms.

While we monitor and prepare for tropical storms over days, tsunamis present the most risk to society. The 2004 tsunami off the coast of Indonesia caused 230,000 deaths on the island of Sumatra and resulted in thousand more around the Indian Ocean and South Pacific. Tsunamis are triggered by major movements in the seabed that rapidly transfer kinetic energy (movement) to water. The 2004 tsunami was triggered by an undersea earthquake, as are most tsunamis. The wave can travel hundreds of miles per hour and retain magnitude until disrupted by a land mass. The 2004 tsunami produced a surge as high as 80 feet in Aceh, Indonesia and was observed even on the west coast of India as well as the west coast of South America. Unlike tropical storms, the risk factors for tsunamis are not based on climate or season. They occur from seismic activity, putting coasts around the Pacific Ocean's "Ring of Fire" at higher risk. Japan has experienced numerous earthquakes and tsunamis over its history, impacting both laws and culture. Much of the damage of the 2011 Fukushima nuclear meltdown was caused by the flooding of the nuclear reactor by a tsunami resulting from a seabed earthquake.

Rivers collect water from many land sources and transport it ultimately to the ocean, although in a few places to isolated saline lakes and seas. Streams transfer water to small upstream rivers, which transport water to larger rivers or lakes. Lakes temporarily store water, typically bound for the ocean. Lakes are fed by rivers and streams and are typically drained by rivers. The residence time for water in lakes can be as short as hours to as long as two hundred years for Lake Superior or ten thousand years for Antarctica's Lake Vostok. The residence or retention time is calculated from a ratio of the lake volume to the flow volume. As we will discuss in Chapters 4 and 5, rivers and lakes are the backbone for most of society's development of communities, villages, cities, and even nations. Rivers and lakes traditionally provided drinking water. Rivers and lakes provide food from fish. Rivers were traditionally a primary mechanism for transporting goods and people. Rivers obey one physical force: gravity, which dictates their pathways and flow rates. The circuitous route of rivers is created by the action of gravity on water in response to physical diversions created by impervious materials such as rocks. Rivers widen and spread out when the land is flat. Heavy rains or rapid snow melts increase water flow causing rivers to rise, potentially dislocating softer materials and creating new channels for the river. When rivers top their banks and flood, whether from snow melt or excessive rain, the surrounding land is replenished, supporting plant growth directly and animal growth indirectly. As we will delve into in the next chapter, Ancient Egyptian culture developed surrounding the productive agriculture land created by regular Nile River flooding.

Rivers and lakes are much lower in salinity than the ocean. River salinity increases in the deltas where they mix with ocean water. The mineral and nutrient content in rivers and lakes are strongly dependent on the activity surrounding them and upstream from them. As with the ocean, the temperatures vary strongly by location and season. A big difference is that freshwater freezes much easier than salt water, so, in cold climates, lakes freeze. Depending on the water flow rate, pockets of rivers can also freeze. As we discussed in Chapter 1, water has unique physical properties allowing ice to float on water. Therefore, a frozen layer on the surface of a lake prevents the subsurface water of the lake from freezing and protects the fish and fauna community. Depending on the porosity of the lakebed, lakes can be substantial sources of ground water recharge. Basically, water seeps through the soils and rocks, filling underground aquifers. The age of water in underground aquifers is an area of significant research and important implications for nature and society. For example, the Ogallala Aquifer under the Great Plains in the U.S. provides more than three-quarters of the region's drinking water and is used heavily in irrigation. It is being rapidly depleted by human extraction and will require thousands of years to naturally recharge.

While we often think of aquatic life as separated between freshwater and saline species, rivers and lakes provide essential support for the lifecycle of many saline species. Most well-known are salmon, but there are many other fish that can change the way they manage salt via osmoregulation — collectively known as euryhaline fish (includes bull sharks, gulf sturgeon, barramundi, and Atlantic stingrays, to name a few). Salmon return to the same freshwater location to spawn, lay their eggs, and die. The rivers and lakes could not provide enough food to support the full adult population, thus they return to the ocean. The streams, rivers, and lakes provide protection for the eggs to hatch and find shelter from predators. Thus, rivers enable the ocean fish population to thrive.

While we think of rivers and lakes as natural water bodies, artificial lakes and canals have been constructed for millennia. Canals tend to be straighter than rivers, but constructed lakes become identical to natural lakes within a few years. Aquatic populations in artificial water bodies, too, become essentially identical to their natural counterparts. Why do societies construct water systems? There are several reasons. Lakes may be created as reservoirs for municipal water supplies or on farms for irrigation in the dry season or animal watering. Canals are typically constructed to connect water bodies for the transport and trade of goods. With either, constructed water bodies offer the potential for productive work. Hydropower plants produce electricity at river dams, and the artificial lake behind the dam serves both a water and energy function. Mills have been constructed on rivers for centuries to provide the energy for industrial activity long before the development of electric power. There are other types of water bodies that seem similar to lakes and rivers but represent distinct natural systems. One of the authors was born a stone's throw from the East River in Manhattan. This water body has famous bridges such as the Brooklyn Bridge and the 59th Street Bridge spanning it. Few New Yorkers realize that, actually, it is not a river. The East River is an estuary. An estuary looks like a river but serves a different function. Rather than transport water from the highlands to the sea, estuaries connect two bodies of water. On the west side of Manhattan, the Hudson River is fresh, and water flows from north to south. The (misnamed) East River is saline, and water doesn't flow in either direction. Instead it connects the Long Island Sound with the Upper New York Bay, both of which are connected to the Atlantic Ocean. So next time someone insists that the East River is a river, tell them you have a bridge for sale.

Wetlands are natural water bodies that are similar to lakes but play a distinct role in nature. Many refer to wetland swamps with negative connotations, and countless swamps have been drained throughout human civilization. Considering their pivotal role in cleaning water, we should be restoring wetlands rather than draining swamps. In wet seasons, wetlands can appear similar to lakes. In dry seasons, wetlands seem to become pastures. Wetlands are shallow water bodies that serve as a natural system for recycling nutrients from agriculture. These ecosystems are home to amphibians such as frogs, protecting them from larger predators that live in lakes and rivers. The cycle of wetlands is that they may start as lakes, but over time, sedimentation of silt and other materials reduces their depth. Eventually wetlands may become pastures. Beavers are frequently integral to the lifecycle of wetlands. Their dam building causes the silt build-up and precipitates the transition from lake to wetland. Because of this type of effect, beavers are second only to humans in their ability to cause macroscopic changes to the environment.

Water in geology

We are all more or less familiar with the behavior and location of water on the surface of Earth. As complex as those systems seem, subsurface water exists in a far more complex environment. In a lake, virtually every drop of water touches only other water, and only interacts with other solid substances at lake bottoms and the like. Subsurface water, in contrast, is often in intimate contact with rock, sand, shale, and a range of other substances. The complexity is so broad that hydrogeology is a subject on its own. In the context of human use, we are primarily interested in subsurface water that is within a few hundred feet of the surface. Some basic questions about subsurface water include: How does it get down there? Where does it go? What does it do down there? In what condition is it? And when and how does it come back to the surface? Each of these questions is rich and can genuinely be answered with "it depends." Most water gets below the surface by infiltrating through porous rock, soil, and sand. Some water may be transferred from deep underground. This water can be transferred from the bowels of the planet upward at fault lines or during seismic activity. Here we will focus on the water that infiltrates from the surface. Near the surface, water content or soil moisture can range from essentially dry in deserts, to saturated around wetlands. Within a few feet of the surface, the mixture of water and air creates an inviting environment for life, from microorganisms, to insects, to the roots of plants. Water acts as the transport fluid allowing roots to take up nutrients from the soil.

As on the surface, if there is a pathway, water will continue to flow downward, driven by gravity. Such penetration requires porous soils, sands, or porous rocks such as sandstone. If there is a band of impervious rock under the surface, such as granite, its lack of porosity blocks the water from penetrating further (except through larger cracks), and the water

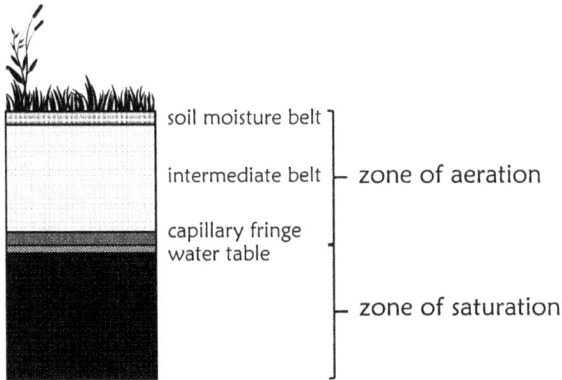

soil moisture belt ⎤
intermediate belt ⎦ zone of aeration

capillary fringe
water table

zone of saturation

Figure 3.3. Underground water is divided into the zone of aeration and the zone of saturation, with the water table at the upper surface of the latter. The zone of aeration contains a belt of soil moisture from which plants draw their water.

tends to pool on top of that layer. This groundwater is what we tap into when we dig a well. As we move down toward the center of the earth, there could be numerous regions where water was able to infiltrate until it collects above an impervious surface. When talking about sub-surface water, we split things into two main zones: the aeration zone and the saturation zone (Figure 3.3). The zone of aeration is a layer of materials (soil, rock, etc.) in which the various pores and cracks are filled with both air and water. Rainwater and snowmelt infiltrating from above represent the primary source of water in this zone. Some water also rises up from below via capillarity (see Chapter 1). Exactly how thick the zone of aeration is will depend on the specifics of the location. Arid climate, high elevation, and highly permeable materials all tend to increase the thickness of the aeration zone. The porosity of this zone is also an important factor in determining how rapidly groundwater can be recharged. Below the zone of aeration sits a layer of permeable rock and soil in which all the pores and cracks are filled with water, all sitting on top of the impermeable layer underneath. The top of the zone of saturation is known as the water table.

Interestingly, whereas impervious rock traps water from penetrating deeper, the inverse story applies to natural gas. When natural gas seeps up from deeper underground, it is trapped when it hits an impervious surface above it. Therefore, we frequently find deep groundwater located close to natural gas resources. The natural gas can be intercalated (enmeshed in) shale rock. With new technology, we extract this shale gas by fracturing the rock with a mixture of water, sand, and chemicals in a process called hydraulic fracturing, or "hydrofracturing," or, most colloquially, "fracking." When we extract natural gas or oil, we actually recover more water than fossil fuels by volume. Some of that recovered water is the "frac fluid" that was originally used to crack the shale, but we also recover lots of other water that was located close to the buried fossil fuel resource. This water is called "produced water." To the fossil extraction industry, produced water is considered a costly inconvenience. A significant fraction is managed by reinjecting the produced water deep underground. Normally, you reinject away from the extraction well to avoid re-extracting the same water, in what would otherwise be a rather silly exercise. Reinjection has caused induced seismicity, especially in the region around Oklahoma — leading some localities to ban the practice altogether. In arid regions, produced water is viewed as a potential resource for irrigation. Produced water quality is dependent on its origins. Some water could be a little brackish (a few thousand ppm in salt) and relatively easy to purify enough for irrigation. In other cases, it's more complicated — and environmentally challenging.

Coal has a different relationship with water. When subsurface coal is exposed to water, it starts generating natural gas. This energy source, called coal bed methane, is generated by anaerobic bacterial reactions with the coal. If you look at the chemistry of these reactions, they also generate carbon dioxide. Methane (natural gas) is not very soluble in water, but carbon dioxide is. Therefore, produced water from coal bed methane has a signature water quality. It is always high in dissolved carbon dioxide. We are all familiar with dissolved carbon dioxide. It converts into bicarbonate and makes the water fizzy and tart (acidic). We drink it as sparkling water or club soda — don't drink the coal bed methane water, though.

There are cases where underground water has high relevance to energy beyond fossil fuels as well. From fault lines, tectonic plate movement, and volcanic activity, there are many places where heat from within Earth rises up near the surface. When porosity in the sub-surface allows water to make contact with hot rock or lava, the water heats up. Sometimes, even though it can get very hot, it remains in the liquid state. If this water finds its way to the surface by natural or anthropogenic means, we have hot springs. Societies have been attracted to hot springs for millennia. They are considered a place to restore health and relax the mind. Superheated water will eventually boil, and the steam can force a large volume of water to spray out of the ground. One afternoon, late in the summer of 1870, members of the Washburn-Langford-Doane Expedition ventured down the Firehole River in Wyoming, in what is now Yellowstone National Park. These explorers were amazed to discover great plumes of boiling water bursting into the air. The first geyser they saw seemed to erupt with regularity, leading them to name it "Old Faithful." In such geysers, the water infiltrates the hot rock at a relatively constant rate, building pressure, which eventually is released in these eruptions. Hot groundwater is a viable renewable energy resource. We can use it to heat a home or building by recirculating hot water from underground. More frequently, and on a larger scale, we use the steam to drive a turbine and generate electricity. Geothermal energy is considered both renewable and "baseload" energy. Baseload means that the electricity generation is stable, unlike sources such as solar and wind, which can be highly variable. A downside of geothermal energy is that it is only viable in areas with high seismic activity. Going beyond conventional geothermal, as you go deeper in the earth, the temperature gets hotter and hotter. This is true even away from seismic zones. There has been significant research examining the feasibility of injecting water from the surface, letting the superhot rock generate steam and, thereby, electricity. So-called "enhanced" geothermal commercialization has been plagued by the high costs of digging deep wells and providing large volumes of water.

Beyond generating renewable energy, groundwater is also seen as a potential way to deal with carbon dioxide emissions that affect our climate. Deep aquifers located thousands of feet below the surface can be colossal and essentially isolated from the surface. This deep water tends to be highly saline. As we mentioned previously, it is easy to dissolve carbon dioxide in water — and even easier in saline water, so perhaps this is a place we can stick a bunch of it. As we inject carbon dioxide into the deep aquifer, we will need to extract enough of the saline water to provide space for the carbon dioxide and avoid too much pressure build-up. Such extracted water would represent a significant treatment or disposal challenge, so many researchers are developing methods specially designed to treat highly saline, extracted water.

Water shapes the environment

The fact that water shapes the environment is among the most obvious aspects of this entire book. A common imagery of how water shapes the underground environment is in accessible caverns and caves. The humidity is close to 100%, the temperature is stable in the 50s (Fahrenheit), and the water is highly concentrated with minerals. Dripping water creates stalactites and stalagmites. These structures require centuries to form and are beautiful demonstrations of water shaping the underground environment.

Without water, there is no life. Without water, there is no flow of minerals from the mountains to the sea. The erosion properties of water erode and smooth surfaces over time. Minerals that precipitate from water eventually become sedimentary rocks and cover about three-quarters of Earth's surface. We think of rivers as channels, but think bigger — the Grand Canyon was etched out by the Colorado River. Without the deposition of silt by the Hudson River, there is no Manhattan. The Amazon region in the northern part of South America would look like the Sahara Desert were it not for the river. Considering physics and chemistry, the ocean buffers most of the heat on Earth's surface and absorbs most of the carbon dioxide from the atmosphere. To say what Earth would look like without water is beyond consideration. Without water, Earth is the Moon — no rich spectrum of colors, no dynamism, no fun, no life, and no book about water.

04 Water is Society

A brief history of water and civilization

Human civilization throughout history can be viewed through the lens of water. It has dictated the rise and fall of many great cultures. A full accounting of our connection with water through history would be an impossible task; here we aim to catch some key developments and to give a flavor for the ascendant role of water.

We can begin the story of our relationship with water at the beginning of the story of human history. Our ancestors existed as hunter-gatherers wandering from one seasonal water hole to the next side-by-side with animal herds. Water was found, and it was used where it was found. This dynamic began to shift around 10,000 years ago. As glaciers receded back to their northern havens in response to planetary warming, temperate forests regained a foothold. Bison, elk, and other large mammals followed the tundra grass and moss north, leading some early humans to abandon the hunt, settle down, and adopt farming life. For the settled farmer, water management became synonymous with survival. Barley and wheat cultivation in the Fertile Crescent initially relied on (generally) bountiful rain. Vulnerability due to these intermittent rains led farmers to some of human's first efforts to take control of water. These rather transient hillside settlements eventually gave way to early walled farming and trading settlements like the ancient city of Jericho in the Jordan River valley, fed by the biblical Elisha's Spring, and to farms on semi-arid river plains that boosted natural rainfall with irrigation. Lands fed by irrigation and fertile silt from the Tigris and Euphrates Rivers in Mesopotamia produced food surpluses, which in turn led to unprecedented population growth. Until the Industrial Revolution, population size was limited by the availability of water. River-based irrigation civilizations emerged, leveraging wooden and reed rafts for communication and trade on arterial waterways, with concentrated political power. Elements of modern society developed, such as writing and taxation. History's first great empires were born.

Agrarian civilizations took hold around the Tigris/Euphrates, Nile, Indus, and Yellow Rivers. Each shared attributes of hereditary authoritarian rule over a hierarchical society led by an elite class. Power was derived from water, principally through its interrelation with food production. Centralized — often brutal — control over masses of peasant laborers enabled the construction and maintenance of complex irrigation networks comprised of canals, levees, storage dams, and the like to ensure adequate supplies of water and fertile silt while at the same time mitigating the devastation of floods and offering routes for military deployment, commerce, and regional communication. Water has traditionally been the most efficient means to transport heavy goods. Whenever natural or political forces interfered substantially with the flow of water, the resulting decreases in crop yields decimated food supplies and threatened the entire social construct.

"Mesopotamia" stems from the Greek for "land between the rivers," in this case the land between the Tigris and Euphrates — rivers with historically challenging water behavior characterized by erratic flooding and substantial, sudden changes of course that have the potential to leave a farm or even an entire community isolated from its surroundings or without water. Ancient civilizations in the Fertile Crescent achieved success by overcoming these challenges with water engineering. Mesopotamians implemented ditches and sluices to assist with water drainage, developed means of lifting water to flood farmland with both water and precious silt, and constructed large reservoir dams to store water to last through extended dry periods. The centrality of water in Mesopotamian culture is epitomized by the mythological Enki, revered Sumerian god of water (later known as Ea in the following Mesopotamian Akkadian and Babylonian mythologies). Regional mythology also has a focal flood myth, strikingly similar to the later story of Noah in the book of Genesis, which reveals a recognition of water as both a giver of life and a bringer of catastrophe.

Recognition of the growing importance of water to Mesopotamian civilization led strategists to engineer its presence or absence as a powerful weapon of war and means of maintaining loyalty. Through time, the center of Mesopotamian civilization kept moving progressively further upstream along the Tigris and Euphrates basin, first with the city of Sumer near the Persian Gulf, then Akkad and later Babylon near the modern city of Baghdad, followed by Assur in what is now northern Iraq. In this way the centralized leadership could better protect the source of water that fed and transported their people. Failure to control one's own source of water could result in disastrous consequences. Around 2450 BC, Umma and Lagash, neighboring ancient Sumerian city-states, erupted into a violent dispute over borders. This conflict was the world's first water war. Each side dug canals both to provide themselves with independent sources of water and to divert water away from their enemy, to devastating effect. Centuries later, the famed Babylonian leader Hammurabi maintained a tight grip on power over a collection of city-states using precisely the same strategy.

The Sumerians, likely the first to develop a massive hydraulic society through water engineering on an unprecedented scale, eventually faced their downfall as a result of that very same movement and use of water. Water engineering represents the first major infrastructure projects of society. Sumerian farming efforts to feed an ever-growing population included not only large-scale irrigation of the arid land, but also deforestation. Soils in Sumer suffered from poor drainage, and when heavily farmed without the restorative benefit of trees, ruinous salinization took hold. Wheat yields were no longer sufficient to feed Sumer, and shifting to more salt-tolerant barley only served as a temporary solution. Population plummeted, and the political balance shifted. Akkadian civilization, with a power base further to the north, assumed power. The Akkadian Empire controlled Mesopotamia, the Levant, and parts of Anatolia, sending military expeditions into the Arabian Peninsula. Although historical records are inconclusive, it is probable that the Akkadians (and subsequently the Guitans) also fell victim of water's fickle ways. A climate shift to drier and windier conditions decimated their food supply, splintering the society into two major Akkadian-speaking nations: Assyria in the north, and, a few centuries later, Babylonia in the south.

Two water-related technology breakthroughs, pioneered in large part by the Assyrians, led to further power shifts in Mesopotamia. The first was a military advancement. Although some iron artifacts have been discovered dating deep into antiquity, cast bronze had dominated metallurgy in Mesopotamia for nearly 2,000 years. Iron ore was originally smelted in charcoal-fueled, bellowed furnaces called bloomeries. Carbon monoxide produced by the charcoal reduced iron oxide from the ore to iron metal. Metalworkers would laboriously beat and fold the resulting mass, called a "bloom," to force out the molten slag and produce wrought iron, which is a rather soft alloy with no clear advantage over bronze. Assyrian smiths discovered that wrought iron could be transformed into a far harder material by heating the finished piece in a bed of charcoal and then quenching it — in water. Quenching turned the outer layers of the metal into steel with an inner core of less brittle iron. This new metal produced hard weapons that had a distinct advantage on the battlefield, marking the end of the Bronze Age. The second technology breakthrough led to a revolution in obtaining clean urban drinking water: the qanāt. "Qanāt" is Arabic for "channel," which points to its identity. These were gently sloped underground channels used to move water from an aquifer or spring into a settlement. Being enclosed beneath the earth, evaporation in the hot, arid Mesopotamian climate was dramatically reduced relative to transport using open canals and ditches. Moreover, the presence of hills on the surface would not impede the flow. Residents of ancient cities finally had a reliable, secure source of freshwater not only for drinking, but also for irrigation. Jerusalem, the City of David, rose to prominence in the Levant using a similar technique to transport water from the Gihon Spring (Fountain for the Virgin) in the Kidron Valley through underground tunnels that still exist today. The spring is located outside the walls of the ancient city, and therefore it represented a major military weakness in the absence of a means to get the water safely within the walls. The existence of these tunnels was a closely guarded secret for many years since they provided effective resistance to sieges of the city including, ironically enough, by the Assyrians. Qanāts were such an effective means of moving water that tens of thousands of them remain in use in cities such as Tehran to this day.

Persian civilization, launched emphatically by Cyrus the Great's creation of the Achaemenid Empire, represented the largest empire in human history, spanning at its peak from the Balkans in Europe to the Indus Valley. Cyrus, and later Persian rulers such as Darius and Xerxes, adopted numerous advanced hydraulic methods to strengthen and expand power. Engineers established a grid pattern of canals, thereby expanding irrigation. They dredged rivers to actively remove silt; they addressed the salinization challenge that was the downfall of the Sumerians by planting weeds during fallow seasons that would lower the water table; and they introduced a stunning piece of water technology called a noria (Figure 4.1). A noria can be viewed as an ancestral relative of the waterwheel; it draws

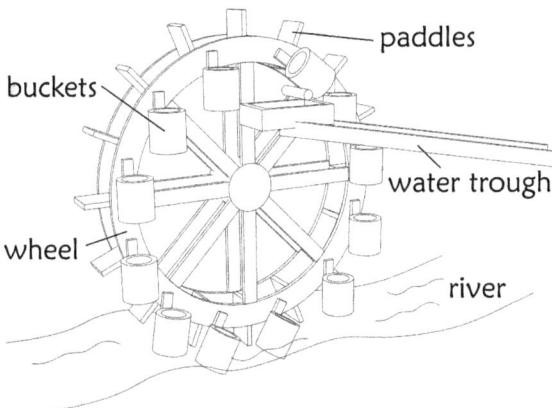

Figure 4.1.   Schematic of a noria.

its energy from the flow of a river. Designed to lift water to an elevated aqueduct for subsequent distribution by gravity, noria consist of large, narrow wheels whose rims contain a series of water-carrying containers that fill up as they pass through the river and empty at the apex of the wheel's circle. The noria was later adopted by societies further to the west, along the Nile River. Those ancient Egyptians were no slouches themselves when it came to water technology and management. We will turn to them next.

Ancient Egyptians, like their neighbors in Mesopotamia and the Levant, occupied a semi-arid land that became phenomenally productive for agriculture thanks to irrigation and fertile silt from an arterial river. Water was central to maintaining power, and the pharaohs implemented technologies such as the noria in recognition of water's supremacy. Other important technological contributions included nilometers and shadoofs. A nilometer (Figure 4.2) was a step-like structure for measuring the Nile's level as well as its clarity during the summer flood season. If the nilometer revealed a low flood level, decreased agricultural productivity was forecast; if it showed high levels, inundation and destruction could be

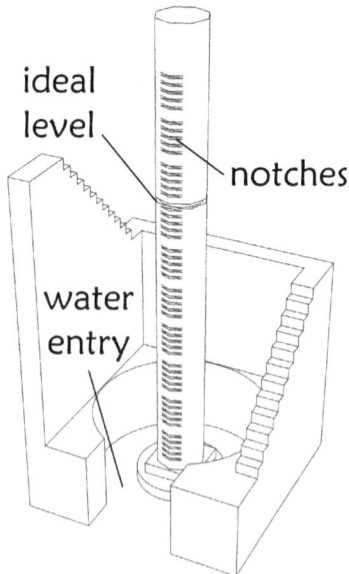

Figure 4.2.   Schematic of a nilometer.

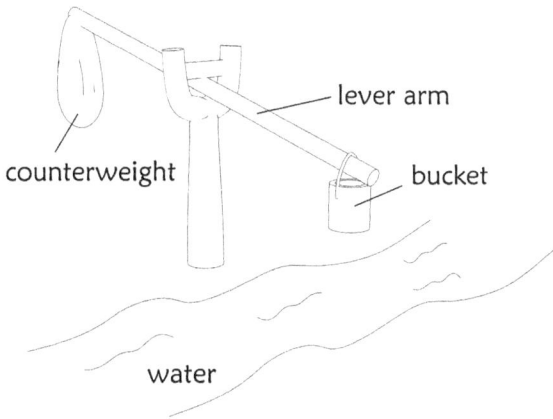

Figure 4.3.   Schematic of a shadoof.

expected to ensue. Nilometers carried a specific mark that indicated the ideal flood level that would provide the fields with just the right amount of silt. Egyptian priests garnered much of their power from their ability to predict the volume of the coming inundation, which surely seemed magical to the uneducated masses. Prediction of flood levels also played an important administrative role, since flood quality determined the taxes to be paid by farmers. Shadoofs, in contrast, served a function similar to the noria, namely, irrigation. A shadoof (Figure 4.3) is essentially a means to reduce the effort needed to lift water. It is comprised of a bucket secured to a lever arm and a tall fulcrum pole stuck in the ground. This simple construction was so useful that there are still regions of Asia, Africa, and Europe where one can find shadoofs in use today.

Shallow waters around the Nile Delta served as home to a deceptively important aquatic flowering plant called Cyperus papyrus. If one removes the outer rind from the stems of this tall plant and then harvests the fibrous inner pithy material, one has the makings of one of history's central inventions. Cutting the pith into strips and placing them, overlapping slightly, side-by-side, and repeating the process with a second layer with the strips aligned perpendicular to the first layer one can construct a flat sheet of material. Soaking this sheet in water and then hammering the sheets together produces papyrus paper, the substrate for countless historical writings — another notch in water's belt as the world's most essential material.

Ancient Egypt's fortunes rose and fell with water. The Old Kingdom, a great power for five centuries, collapsed in large part as the result of a climatic shift to a dry period that rendered Egypt's once abundant agricultural production inadequate. Regional governors no longer relied on the pharaoh in times of crisis, food shortages escalated into full-blown famines, and civil wars spread across the land. This First Intermediate Period lasted nearly 200 years before abundant floods returned, restoring the central government to power in the Middle Kingdom, stimulating a resurgence of literature, art, and mammoth building ventures. When another series of droughts weakened the pharaohs again, a western Asian civilization called the Hyksos invaded and settled in the eastern Nile Delta, initiating the Second Intermediate Period of ancient Egypt. A new era of nourishing floods 125 years later ushered in the New Kingdom whose pharaohs established a period of unprecedented prosperity. Egypt secured its borders and, in large part as a product of foreign sea trade, built strong diplomatic ties with neighboring civilizations. One ruler of the New Kingdom, in particular, oversaw major water innovations: Pharaoh Necho II. During a campaign in Syria, Necho II started an ambitious project to dig a navigable canal from a branch of the Nile to the Red Sea — a precursor of today's Suez Canal. Necho also broke with Egyptian tradition of fearing the sea and formed a navy comprised of Greek sailors, which operated along the Red Sea and Mediterranean coasts. While the Egyptians were dominating the Nile region and the Mesopotamian cultures the Tigris and Euphrates, further to the east another great Old World hydraulic civilization left its mark on human history.

The Indus Valley civilization flourished in a region where Pakistan and parts of India sit today. Although there were indications in the middle of the 19th century that the Indus peoples had existed, it was surprisingly not until nearly a century later that archeologists caught onto the fact that the region around the Indus was home to one of the great ancient civilizations. Unlike their counterparts to the west, the Indus (Harappan) people thrived in a monsoonal habitat, meaning the annual delivery of water was heavily concentrated in a short period during the summer. (Their water supply was supplemented by snowmelt from the Himalayan and Hindu Kush Mountains, which remains a vital source of water for much of Asia today.) Survival meant water had to be stored during the wet season for later release during the dry months, and the Indus engineers successfully implemented a sophisticated system of reservoirs to this end. Even more impressively, the Harappans were nothing less than genius hydraulic technologists, in some cases developing technologies millennia before they appeared in other civilizations. We tend to think of the ancient Romans as the originators of most urban hydraulics (we will get to them in a bit), but preceding ancient Rome by 2,000 years was the communal Great Bath at Mohenjo-daro, one of the largest settlements of the Indus Valley civilization, and also home to enough wells to provide almost every dwelling with its own private water source. Harappan engineers also installed underground sewer systems constructed carefully from brick, demonstrating a recognition of the hazards associated with disease-spreading waste. Most shocking of all, Indus cities contained indoor toilets. Keep in mind that this is 4,000 years before the sanitary revolution (discussed later in this section). Despite all this remarkable modernism, Harappan civilization, like its neighbors, ultimately met its demise at the hands of water. Archeologists point to catastrophic floods, which rerouted great rivers, as the likely driving force behind the civilization's downfall. Indeed, similar water-related climate change fates, whether they be floods or droughts, are associated with many other ancient cultures such as the Pueblo societies, the Mayans, and the Hohokam in pre-Incan Peru.

The Old World, however, was not the only setting for early hydraulic civilizations. Ancient China also boasts a rich history in vast societies centered on water. The Yellow River, originating in the Bayan Har Mountains in western China and crossing the northern region of the country, was the birthplace of ancient Chinese civilization (see Figure 4.4). Soil in the north is dominated by loess, a yellowish, silty sediment formed by accumulation of dust blown in from the Gobi Desert, and the Yellow

Figure 4.4.   Major rivers in China.

River, which collects vast quantities of loess, is the world's siltiest river. Shifting deposits of this silt leads to sporadic massive flooding, which has killed literally millions of people and produced numerous social upheavals throughout Chinese history. Flooding, and efforts to mitigate its devastation, is so integrated into Chinese culture that numerous leaders in China have been hydraulic engineers. Although historians continue to debate its historical accuracy, the attributed founding father of the Yellow River civilization and initiator of the Xia Dynasty was Yu the Great — one such water engineer. Social stability in China has almost always depended on bringing water under control, pacifying and channeling it. Culturally, a leader who manages to do this successfully is considered a virtuous person and is bestowed with the right to rule. Yu devised a system of flood controls that, rather than directly damming the Yellow River, implemented irrigation canals that relieved floodwater into fields and dredged riverbeds to remove silt deposits. This approach is aligned with a general philosophy, Daoist-leaning, of designing waterworks that allow water to flow and utilize diversion weirs and the natural ecosystem. Such a philosophy is in stark contrast to the more Confucian approach of forcefully manipulating water with dikes and dams. These two diverging schemes have traded influence over hydraulic designs through much of Chinese history.

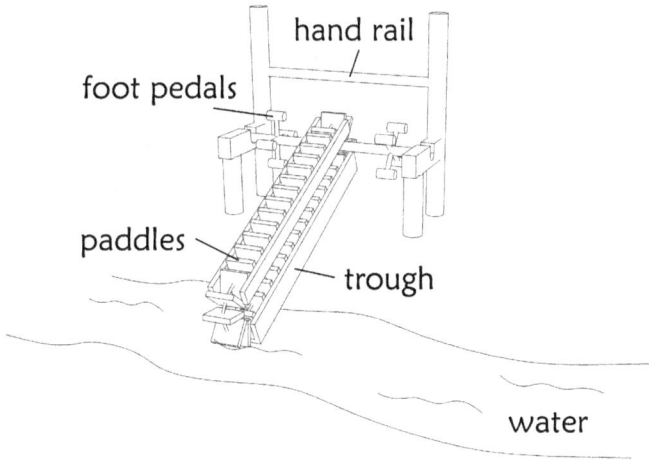

Figure 4.5.   Schematic of a foot-treadle chain pump.

Ancient Chinese engineers introduced a number of water technologies with lasting impact on the region. They invented deep brine wells, used to produce precious salt, which employed a natural resource to form rudimentary plumbing for transporting water: bamboo. Pipes constructed of bamboo were also used in rice farming and even as water mains in early cities. Another deeply influential technology was the foot-treadle chain pump (Figure 4.5). These devices were powered by one or two laborers who used a stepping motion to turn a string of wooden paddles along a trough. The paddles scooped water into the trough and lifted it onto a field or into a channel for irrigation and other use. Treadle chain pumps can still be found in many regions of rural China.

Waterwheel machines, like the noria in North Africa and the Middle East, also appeared in ancient China. Chinese waterwheels most likely originated independently from those further to the west, as early Chinese systems were generally horizontally oriented; regardless of their origin, the Chinese were indisputably world leaders in harnessing power from waterwheels. By the Han dynasty, around the 1st century AD, waterwheels milled grain and hulled rice and sifted flour, waterwheel bellows forged iron ore into cast iron, and waterwheel trip hammers crushed that ore and pounded that iron. Another application for waterwheel power was in silk making. China's silk production was the foundational trade product on the Silk Road(s), which themselves followed paths through Asia dotted with natural water sources to supply the merchant caravans.

Another major water contribution stemming from the Han dynasty was the Lingqu (Magic) Canal, the world's first transport contour canal. A contour canal is a meandering navigable water course that follows the contour line of the land to avoid engineering structures such as tunnels through hills or embankments over lower ground. The Lingqu Canal created a waterway connection through the mountain chain that divided the Xiang River (which flows into the Yangtze) and the Pearl River Delta to the southwest, permitting unprecedented transport over a continuous distance longer than 1,000 miles. Benefits afforded by the canal helped the Han maintain relative stability for more than four centuries. Several hundred years later, the Chinese constructed another canal, this time at a far grander scale, representing one of the true transformational events in Chinese history. A labor force even larger than that needed to construct the Great Wall was employed, digging the waterway and building a collection of 60 bridges and 24 locks to navigate changes in elevation. The so-called Grand Canal was completed in the early 7th century AD, and even today it remains the longest artificial waterway on Earth, spanning a distance equivalent to that between the cities of Chicago and Miami. The Grand Canal linked the northern Yellow and southern Yangtze river systems for the first time, merging the benefits of the dry north with its rich loess and the wet south with comparatively poor soil. China was finally integrated into a militarily defensible nation-state.

Subsequent development was further spurred by another water-related advance: the rice farming revolution. Using terraced paddies managed with sluice gates, norias, treadle chain pumps, bamboo pipes, and dams (as well as a new variety of faster-growing rice imported from Southeast Asia), rice production dramatically expanded and, as always happens when food supplies swell, population exploded. With advances in water engineering, land availability had become the limiting factor in rice production, and the terraces converted hilly terrain into useful flat agricultural land. Urban centers emerged with vibrant entrepreneurial, protoindustrial, and scientific activity, predating Europe's Industrial Revolution by centuries. Iron production, gunpowder, textile manufacturing, and water and mechanical clocks appeared. In the 13th century sitting rulers were challenged by waves of marauding Yuan Mongols that overran the existing dynastic leaders. However, their relative lack of expertise in water engineering and management heightened the impact of droughts and floods and eventually weakened Yuan control over their conquered land. Native Chinese civilization returned to power with the Ming dynasty, which oversaw another great period of stability and prosperity. The Ming were master shipbuilders, and they amassed fleets comprised of thousands of ships, making them a dominant naval power in the early 15th century. Admiral Cheng Ho (Zheng He) led a massive fleet that reigned over the Indian Ocean. One can only imagine how history would have been different had this momentum been sustained. Within China, a major project to repair and expand the Grand Canal was completed around this time, including the masterful Heaven Well Locks, which split the flow of several rivers and enabled water managers to regulate seasonal flows with unprecedented precision. Thereby, the Grand Canal became an all-season waterway, and sea transport become somewhat redundant. In 1433, the emperor turned his attention inward, limiting Chinese seafaring and restricting contact with foreign peoples, effectively ending the expanding influence spearheaded by Ho and his compatriots. The fledgling merchant class was outmaneuvered by vested agricultural interests. Relative isolation took hold.

Turning our attention back to the west, one can find roots of a different type of civilization, not modeled on centralized, hydraulic authority such as in Mesopotamia, Egypt, the Indus Valley, and China, but rather modeled on a different water perspective: naval power and maritime commerce. Unlike the authoritarian societies, these civilizations, based in the Mediterranean, adopted an emphasis on individual citizen rights and a private-sector economy. On the edges of the great empires in the region, seafaring culture coalesced first in ancient Greece, eventually spreading to form the Roman Empire, followed by republics such as Genoa and Venice. With no major arterial river or broad fertile plains suitable for large-scale agriculture, these societies turned instead to the sea for economic development and protection. The Mediterranean Sea offered easily and safely navigable transport, linked through the Nile Delta into North Africa and through the Bosporus and Dardanelles Straits to the Black Sea and central Asia. Visual landmarks provided sufficient guidance for navigation. Many of these trade routes were pioneered by the Minoans based on the strategically positioned island of Crete.

For centuries the Minoans were a powerful naval and trading influence in the region. Their urban centers, epitomized by Europe's oldest city Knossos, exhibited sophisticated water engineering replete with indoor toilets in the king's palaces and terra-cotta sewer systems under the ground. Minoan dominance was disrupted in large part by a massive volcanic eruption and ensuing tidal wave around 1470 BC, finally giving way to Mycenaen culture on the Greek mainland. City-states, including Sparta and Troy, jostled with Mycenae for power over the next few hundred years, often threatened themselves by iron-weaponed northern invaders. Emerging from this discord were the Phoenicians (Levant), then the Ionian Greek city-states and the Etruscans (Italy). Timber resources in the Levant served as the foundation for a powerful Phoenician navy that reigned for centuries, eventually succumbing to land-based expansion from the east by the Assyrian Empire but founding a descendent naval Mediterranean empire based in Carthage, located in modern Tunisia. Carthaginian civilization would represent an obstacle to the later ambitions of the Romans, which we will get to shortly.

Ionian Greek culture, centered in city-states such as Miletus and Athens, offered up an influential thinker, Thales, the founder of Greek philosophy. Thales believed water was the primal substance of all things (later relegated to being one of four primary elements by Aristotle), demonstrating the centrality of water in human civilization once again. By 600 BC threats from the east resurfaced with the Persian Empire, led by Cyrus and later Darius and Xerxes. Conflict between the Ionian resistance and the Persians culminated in history's first recorded major sea battle, which took place at Salamis in 480 BC. This skirmish saw a vastly outnumbered Greek navy led by Themistocles resoundingly defeat Xerxes' considerable armada thanks to superior ship technology. Greek triremes were quicker, nimbler, and better armed than their opponents. Naval power proved, for perhaps the first but assuredly not the last time, to be an asymmetric advantage enabling a small civilization to successfully face off against a much more formidable, land-based civilization. Following this victory, Athens burgeoned into a rich cultural epicenter, pushing Greece into its classical golden age. A Greek born around this time in the seaport city of Syracuse received credit (still debated among historians) for introducing a new water technology: the screw pump, often referred to with his name as "Archimedes' screw" (Figure 4.6). This machine was used to lift water into an irrigation ditch by turning a screw-shaped core placed within a pipe.

Athens' supremacy faltered in the face of an upstart kingdom led by Philip of Macedon, who cut off food supplies coming from the Black Sea trade route through control over the Bosporus. It was Philip's son,

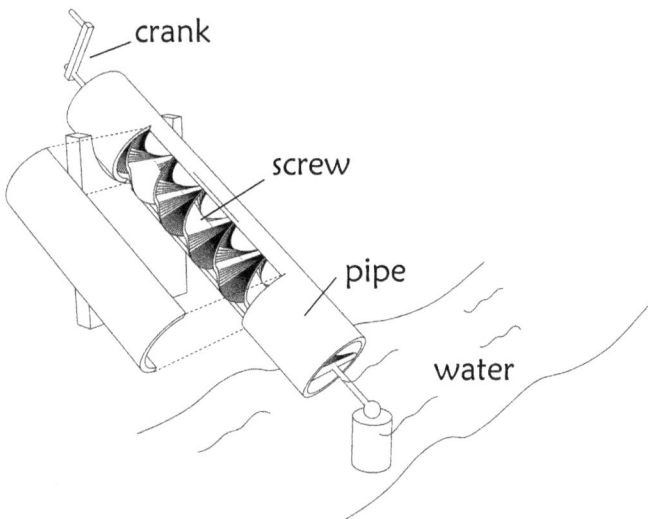

Figure 4.6.   Schematic of Archimedes' screw.

none other than Alexander the Great, however, who would go on to create an unrivaled legacy. In his short life, Alexander led a remarkably successful military campaign spanning 15,000 miles and overseeing an empire spanning from the Nile to the Indus. Athens remained a wealthy city with an active cultural life throughout and following Macedonian rule, eventually being absorbed into the Roman Empire along with the rest of Greece. The Romans, themselves, were preceded by a civilization with a vaunted water tradition. Having mysterious origins, the Etruscans ruled ancient Italy for centuries with Tuscany serving as their heartland. Etruscan hydraulic engineering laid the groundwork for many of the later achievements of Rome. They embanked the Po River against flooding, rendering the territory suitable for prosperous agriculture. These efforts were made all the more impactful upon construction of the magnificent Cloaca Maxima ("great sewer") in the sixth century BC, which effectively drained the boggy, malarial valley in Rome. The resulting dry land served as the foundation for the Forum, the great center of Roman commercial and civic activity. Etruscan engineers constructed the Cloaca Maxima so well that it still functions to this day. It was just a few decades later that the Etruscans were ousted by local Italian tribes, founding ancient Roman civilization.

The Romans, originally a land-based civilization, developed into a sea power, necessitated by their conflict with the Carthaginians. Over the course of three epic Punic Wars, the Romans eventually defeated Carthage; this outward reaching conquest whet Rome's taste for empire, shaping their subsequent history. The waters of the Mediterranean continued to play a central role in Rome's rise, with the city blossoming into a metropolis of around one million that consumed goods imported over the sea from far reaches of the globe. Han China was contemporary with the ancient Romans and engaged in trade with them. Massive grain imports from Egypt and from Central Asia over the Black Sea fed the growing population and armies. Inland, water provided a crucial boundary for ancient Rome. The Rhine-Danube boundary, cutting Europe laterally from modern-day Netherlands to Romania, served as an effective defensive border to the barbarian lands in the north. While the Romans took full advantage of such natural waterways, they were also prolific hydraulic engineers.

Like many other ancient civilizations, the Romans adopted waterwheels, although, unlike the Chinese, the Romans used them almost exclusively for milling grain. As with many things, the Romans went big. They amplified the power of the waterwheel by incorporating gearing, and they constructed massive mill operations that were used for centralized bread production. Speaking of the Romans going big, their public water supply was truly unparalleled. Roman water engineers may not have been the most inventive in ancient history, but they envisioned and implemented a system of unprecedented precision, complexity, and scale. The primary water source was a collection of springs that fed eleven major aqueducts spanning hundreds of miles, fed by gravity. On the way into Rome, these aqueducts processed the water through settling and distribution tanks to remove impurities. To this day we still employ similar approaches as a primary water purification method. In the city, public water supplied more than 1,000 fountains and basins as well as hundreds of public and imperial baths that served as principal institutions of cultural and social life. Wealthy elites even had both hot and cold running water and indoor toilets. Roman engineers fed water with different purity into decorative fountains and functional drinking or bathing sources. This "fit-for-use" strategy was abandoned by subsequent civilizations, including our own today, but there are many who believe it carried intrinsic value and are working to reestablish fit-for-use in modern water systems — we will get to this later in the book. After use in homes and public systems, waste water flowed into a sewer network that carried waste out of the city and into the Tiber River.

In a novel implementation of water power, the Romans invented an early form of hydraulic mining called ground sluicing. Surface streams of water were diverted in order to erode alluvial deposits containing gold. This practice was eventually improved by storing large volumes of water in a reservoir above the area to be mined and then suddenly releasing the water in a powerful wave that removed overburden and exposed bedrock, known as "hushing." Gold veins in the rock were then worked and water power was once again exploited to remove debris. Pliny the Elder provides vibrant descriptions of hushing in the first century AD in *Naturalis Historia,* and Las Médulas in Spain reveals vestiges of the scale of this mining operation with immense badlands terrain resulting from extensive water mining. Hydraulic mining is still used in places today, although its heyday occurred in the mid-19th century during the California Gold Rush.

In addition to utilizing water's physical power, the Romans also developed a revolutionary technology relying on water's chemical prowess. Prior to ancient Roman civilization, robust structures were built using individual stones or bricks, which greatly limited their design and application. The Romans' architectural marvels like the aqueducts, Colosseum, and Pantheon, in contrast, were constructed of a new material: concrete. To produce concrete (Figure 4.7), water is mixed with cement, sand, and gravel, which produces a semi-liquid that workers can shape by pouring into a mold of diverse forms. Concrete solidifies through a chemical process called hydration in which water reacts with the cement, forming a gel and binding the other components together in a robust, waterproof material with the feel of stone. Concrete's influence is hard to

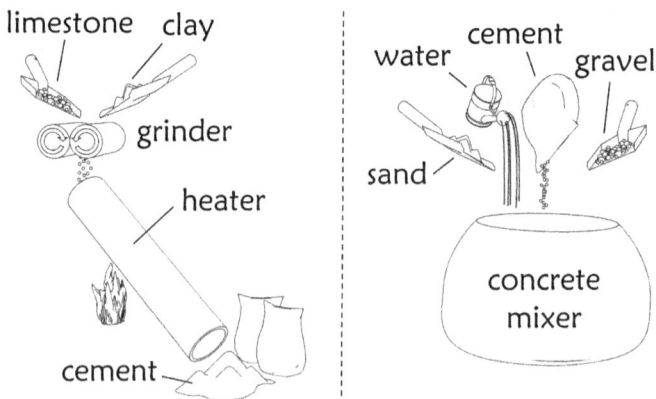

Figure 4.7.   Processes for cement and concrete manufacturing.

overstate; it is the most widely used artificial material today. Amazingly, though, the Pantheon's roof remains the world's largest unreinforced concrete dome.

In much the same way that ancient Egyptian civilization rose and fell with the waters of the Nile's floods and droughts, Roman civilization's health could in large part be tied to its aqueduct-fed water supply. In its prime, ancient Romans had an astonishing 150 gallons of freshwater available for each person every day. This level of water use is comparable to the highest per capita use countries in the modern era such as the United States, Australia, and Canada, and far above today's global average. After over 1,000 years of independence and nearly seven centuries as a great power, the rule of Rome subsided, eventually being overrun by Goths and other barbarians, who themselves were ousted by Huns followed by the Mongols. Aqueducts played a new role in Rome's history when the Byzantine (Eastern) Romans stole into the city through a drained aqueduct, pushing out the barbarians. The Goths attempted to retake Rome by destroying aqueducts as part of a larger siege strategy, but they never regained control of the city. With severely damaged water infrastructure, Rome struggled for nearly a millennium. In the Middle Ages, Romans used wells and cisterns for most of their water, with the poorest residents collecting their water from the dirty Tiber. It was not until the 15th century AD that Rome eventually experienced a revival of power and influence. Pope Nicholas V oversaw restoration of the water supply. The Aqua Virgo aqueduct, originally constructed by Marcus Vipsanius Agrippa in the 1st century BC, was returned to operation by Nicholas. It emptied into a simple basin, the predecessor of the famed Trevi Fountain. Succeeding popes continued this process of restoration, and, as always happens with increased water supplies, population once again soared.

The great ancient hydraulic states and aqueduct-fed societies like the Romans were not, of course, the only civilizations with deep relationships with water. Great civilizations also found ways to flourish even in regions with improbably dry climates. The Arabian Peninsula spawned one such society with resources restricted to a few scattered oases, some underground springs, and the seasonal wadis. Although it sits on the periphery of major rivers like the Nile, Tigris/Euphrates, and Jordan, there is no unifying waterway running through this part of the world. Such extreme water scarcity, probably history's most water-fragile civilization, birthed great esteem for water in both the Arab and Islamic cultures. Tradition demands that no man — nor even his beast — be refused access to drink from another's well. Grains could only effectively be grown along the few small river valleys; otherwise, agriculture was centered on date palms that grew near oases and olive trees along the coasts. Generally speaking, life in Arabia was a subsistence lifestyle. In order for trade to take place across a desert, one needed a means of moving people and goods for large distances without much water. Nature provided just such a means: the camel. With remarkable capacity to store water, camels can go more than a week without water, and they are able to fully rehydrate in a few short minutes, subsisting even on somewhat salty water. Camels could bring Arab merchants to sea waterways, which were traversed on nimble dhows assembled from pieces carried with the caravan. (These were the vehicles of the fictional Sinbad the Sailor.) Disassembling the dhows on the next coast, the caravan could continue.

In the 7th century AD, Muhammad brought together the various regional tribes by controlling the critical oases and the trade routes that relied so heavily upon them. He placed the center of Muslim civilization's burgeoning power first in Medina and later in Mecca. Abu Bakr, Umar, Uthman, and Ali, Muhammad's successors, persistently strengthened the military, rapidly expanding their geographic reach. Because of the limited water — and therefore food — resources, homegrown manpower was limited as well. Expansion, consequently, necessitated conversion and conquest. In a short 15 years, Islamic forces managed to drive a wedge between the powerful Persian and Byzantine empires (themselves weakened in part by a failure to properly maintain the water infrastructure needed to sustain large agricultural production). By the 9th century AD, North and Eastern Africa and Spain to the west and the Indus Valley to the east were absorbed as well. Adopting naval warfare led to a show of force in the Mediterranean and, by taking the Strait of Aden, Indian Ocean trade routes fell under Islamic control. Expansion might very well have continued throughout Europe had a siege of Constantinople been successful. The Byzantines summoned sufficient ingenuity, not surprisingly revolving around water, to withstand the sieges. Engineers used an aqueduct and massive underground cisterns to supply water for long stretches when supplies were largely blockaded. They also introduced a terrifying chemical sea-warfare weapon known as Greek fire. While the precise recipe for this concoction is lost to history, like other ancient incendiary weapons, it was some mixture of sulfur, petroleum, and bitumen compounds. Greek fire would spontaneously ignite and was challenging to quench even when on water. Byzantine defenders would launch Greek fire onto Arab ships, forcing the formidable armada to retreat.

As outward expansion subsided, the few locations with good water supplies swelled into bustling cities. Baghdad emerged as the new center for Islamic society, feeding on the same fertile lands as the ancient Mesopotamians. Engineers restored the canals, qanats, noria, and other waterworks while at the same time constructing new ones. Huge water mills floating on the Tigris River provided Baghdad's daily bread. Importing the technology from China, waterwheel-powered papermaking multiplied (substituting rags for the mulberry tree fibers used in China), eventually proliferating further west into Europe, and, with bookmaking, knowledge spread and flourished.

Across Islamic civilization, however, limited water supplies continued to limit population. Arabs were forced to depend on manpower from the outside, in this case, largely Turks. The imported labor force tended to neglect water system maintenance, eventually allowing silt to clog up the canals, waterlogging soils and instigating salinization. Coupled with a series of low Nile floods and the destruction of irrigation systems in the 13th century AD by Mongol invaders, this had disastrous impact on the society's food yields and ability to sustain a cohesive society. Defeats at sea followed, with control over the Indian Ocean sea-lanes wrested away first by the great Chinese Admiral Cheng Ho and, after the Chinese withdrew from the region, the Portuguese, to whom we will get shortly. The Ottoman Turks themselves emerged as a great regional power, finally taking Constantinople from the Byzantines in the middle of the 15th century AD following a seven-week siege. Turks and Europeans clashed in numerous sea battles, with a serious defeat occurring in the 1571 Battle of Lepanto, considered the world's first major naval battle featuring gunpowder, changing sea combat forever. (Miguel Cervantes, author of *Don Quixote*, was injured in this battle.) With this shift in power, attention swung to Europe.

Two decisive developments in water technology propelled the Western world into a period of global dominance: transoceanic sailing, with long-range cannons evolved from those displayed at Lepanto, and exploitation of waterpower for industry. Having no unifying arterial waterway and being surrounded by the Mediterranean Sea, North and Baltic Seas, and Atlantic Ocean, it may have been inevitable that Europeans would begin to look to the seas for communication and trade. Moreover, without a major river to control, it was not possible for a centralized, authoritarian state to develop; rather, numerous independent forces competed for resources. Northern Europe experienced centuries of power struggles among barbarian and settled societies, with population restricted by limited food production coming from poor natural drainage of its clay soils. With the advent of the moldboard plow, capable of deeper turnover of the soil and in widespread use by the 10th century AD, coupled with crop rotation concepts, areas available for arable cropland expanded dramatically and the population exploded. Inland river trade blossomed, and Europe's wealth grew as never before with urban hubs like Ghent, Bruges, Amsterdam, Antwerp, London, and Paris pulsing with market activity. To the south, Mediterranean city-states paralleled this pattern with Milan, Florence, Venice, and Genoa. Vibrant trade developed between these regions, primarily using Atlantic sea-lanes to move goods once the Strait of Gibraltar was wrested from Islamic control in the late 13th century.

Influenced by powerful market competitive forces, the waterwheel expanded from its traditional use in milling grain to perform a broad

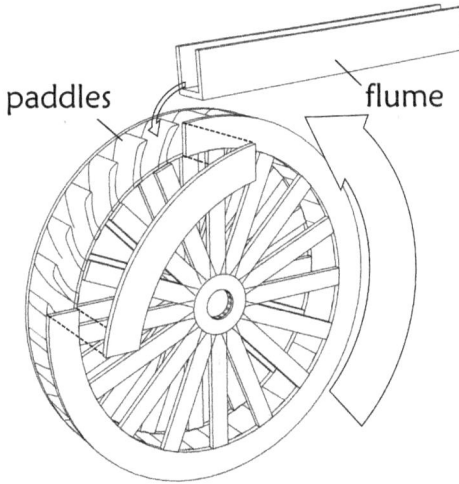

Figure 4.8.   Overshot waterwheel.

suite of mechanical work. One important innovation, trumpeted by none other than Leonardo da Vinci, was the overshot vertical waterwheel configuration (Figure 4.8), which was several times more efficient than the customary undershot design. Waterwheels proliferated into every conceivable stream near inhabited areas in Europe, with hundreds of thousands of them spinning away by the dawn of the Industrial Revolution. Among the most ambitious users were monasteries, which were often situated near rivers and which used waterpower to grind grain, sift flour, hammer cloth and iron, saw wood, and crush olives. Elsewhere in Europe, using the technology from China by way of Baghdad, Damascus, and (Muslim) Spain, water-driven paper mills emerged with massive beaters to pound pulp. With plentiful paper, the ground was laid for the 15th century invention of the printing press — one of history's most influential creations. Textiles and ironmaking were also industries revolutionized by waterpower. When the latter was coupled with gunpowder, it provided the means to arm Europe's vessels and armies.

City-states like Venice had few resources at home, which drove the need for commerce over the seas. Venetian trading ships spread across the Mediterranean and, ultimately, back and forth to Flanders.

The democratic, free-market republic of Venice survived more than 1,000 years. Shipping activity such as this led to a series of innovations in ship technology, producing large sailing ships that were sufficiently sturdy to travel year-round. These ships were the forerunners of those that participated in a world-changing development, centrally involving water, of course: The Voyages of Discovery.

The arrival of transoceanic sailing was spearheaded by the renowned voyages at the tail end of the 15th century AD of Vasco da Gama, John Cabot, and Christopher Columbus. The ability to navigate the world's oceans at will, enabled once explorers unlocked the mysteries of the Atlantic's trade winds and currents, accelerated the interconnectedness of the globe, a process continuing through to today's integrated global society. European countries with comparatively small populations and domestic reach were propelled into positions of world dominance, shifting power balances to emphasize strength at sea over land armies. Portugal, in particular, which introduced the influential and highly capable caravel vessels (Figure 4.9), exerted a disproportionate influence, followed soon thereafter by Spain and then others. In the New World, European muskets and, to a far greater degree, diseases, decimated

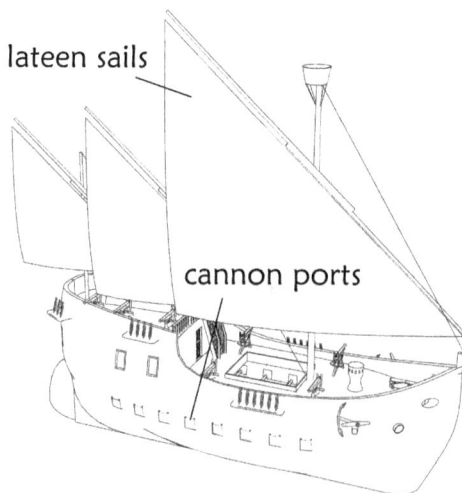

Figure 4.9.   Portuguese caravel.

native populations. Bullion from the New World (pulverized from ore using waterwheel-powered mills) fueled both massive monetary inflation and Spain's rise as a state, catalyzing a period of conflicts that shaped subsequent Western history. As Spain established its New World foothold, benefitting from the infamous papal demarcation line that limited Portugal's territorial claims, the Portuguese pushed around the southern tip of Africa into the Indian Ocean. Their superior cannon-armed caravels easily overpowered any ships that stood in their way, ultimately establishing a presence extending around Africa, across the Indian Ocean, and all the way to the southern Chinese coast — an almost unimaginable feat for a country with a mere one million inhabitants.

Lesser powers England, France, and the Netherlands provided some resistance to Iberian Peninsula power through sea power of their own — largely in the form of state-sanctioned piracy led by the likes of Francis Drake. Privateers eventually weakened the Spanish Armada sufficiently to open opportunities for these upstart nations to gain prominence. The Dutch also demonstrated another form of water mastery: hydraulic engineering with extensive drainage, canals, dams, and dikes to transform a tiny republic sitting, in part, below sea level into a global shipping powerhouse. Like the Portuguese before them, the Dutch reached far and wide with their trade routes and, therefore, their influence. The Dutch East India Company, formed at the onset of the 17th century, wielded great power through its monopoly over trade from Indonesia and Ceylon (modern Sri Lanka). Across the Atlantic, the Dutch formed New Amsterdam, a settlement on the Hudson River for trading in furs from the New World. England soon joined the fray, gaining control over India from the faltering Portuguese and establishing its own footholds in North America. When the French, too, emerged as a colonial power, and the American colonies sought independence from England, wider conflicts ensued. The Napoleonic Wars of the early 1800s demonstrated time and again the decisiveness of sea power in tempering the power of land-based armies. Naval battles during this era marked the last that would be fought by sailing ships constructed of wood. A new water innovation was gathering impetus within England that would, once again, reshape warfare — indeed, it would reshape human history.

Waterwheels played a role in launching the Industrial Revolution, but waterpower in a different form drove much of the explosive development: steam. While the reasons for this development happening in England are many, perhaps the strongest reason was a regional challenge. Hungry for fuel as a result of severe deforestation, the British began exploiting coal mining on a large scale. Building on water technology proven in China and the Mediterranean, insatiable demand for coal led to the construction of thousands of miles of canals through challenging, hilly terrain to efficiently transport the mined product over water. As mines ventured ever deeper below the ground, however, water was also the source of a problem. Ingress from the water table was a burgeoning nuisance. Few mines were situated close enough to rivers to implement waterwheel pumps, so a new power source was needed to pump out the water from deep underground. Although steam power had been theorized dating back to antiquity, it was not until 1712 that Thomas Newcomen built the first successful steam engine. (Earlier models tended to explode.) These were massive, inefficient machines, but they offered a unique means of extracting the problematic water, and they slowly expanded to operate across dozens of English coal mines. Water in gaseous form provided a solution to troublesome water in liquid form. Roughly half a century later, a young Scottish instrument maker dramatically improved on Newcomen's design. James Watt recognized that the efficiency of the engine could be enhanced by avoiding the need to cool the heated cylinder between each stroke, introducing a separate condenser so the cylinder could remain hot throughout operation (Figure 4.10). Watt's new design also carried a far smaller footprint and added rotary motion, extending applicability far beyond just pumping water.

Mechanized cotton textile and cast-iron production in Britain's Midlands region benefitted enormously from steam engines. As more high-quality iron was produced, more steam-powered systems could be fabricated, enabling still more iron production. Factories no longer had

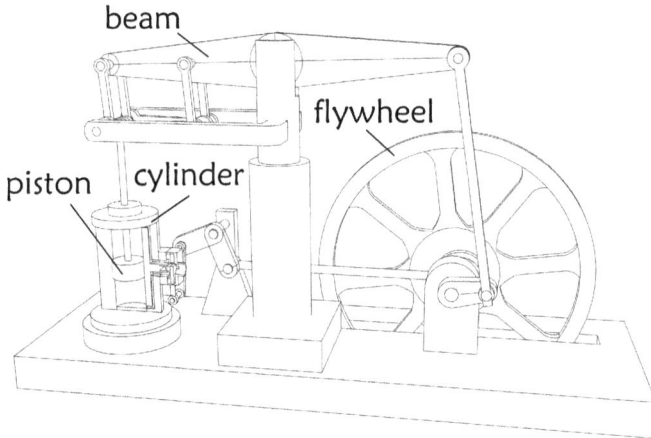

Figure 4.10.   Watt steam engine.

to be sited next to fast-flowing rivers, which were often located in remote areas. Now they could find homes in urban areas where labor was readily accessible, monumentally changing the urban landscape. Steam engines also had direct impact on urban water supplies. It was now possible to lift considerable volumes of water from rivers to supply growing cities for drinking, sanitation, and even firefighting. The old profession of water carrying, in which throngs of laborers toted buckets of water to residents, vanished in short order. Interestingly, waterwheels, in contrast, did not vanish. On the contrary, their use actually expanded because steam engines could supplement the power of waterwheels by providing additional flow. This complementary action, where available, proved more cost effective than coal-powered steam engines alone.

Steam power soon expanded beyond even factories into mobile applications: locomotives, boats, and earthmovers. The latter were capable of dredging land to build giant canals, which we will get to in a moment. Not only did the speed of transportation hasten, but also trade and communication along with it. World economic production grew at a pace unprecedented in human history, and population exploded along with it. Suddenly any country on the planet had the potential to be a supplier of raw materials and a consumer of finished goods. Transoceanic steamers and the 1869 opening of the Suez Canal originally envisioned by Pharaoh Necho II (and the Panama Canal in 1914) unlocked the world to the onslaught of European, and later American, navies and solidified the ascendancy of the west that began with the Voyages of Discovery.

Water- and steam-powered industrialization gave western democracies a powerful kick-start, but a new water challenge soon emerged: water pollution. Long before scientists had an understanding of waterborne diseases, people saw links between water and illness. Ancient Chinese and Greeks boiled water that was suspect, the former buying hot tea or water from street vendors. Alcoholic beverages have represented a sanitary drink dating back to the dawn of human civilization. Another strategy invoked for many years has been the addition of generally available disinfectants such as vinegar. In the 12th century AD, Carthusian monks drilled wells in the Artois province in France where the water jetted out of the ground under its own pressure; these "artesian" wells (named after the province) have been esteemed for their cleanliness ever since. With the rapid urbanization of the Industrial Revolution, however, the condition of water supplies dramatically worsened. Chamber pots were routinely emptied out of the window onto the street or into underground cesspools that seeped into public water supplies. Industrial and hospital waste were generally dumped directly into waterways. The situation worsened, somewhat paradoxically, by the proliferation of the flush toilet. Modern flush toilets, originating in the late 1500s, gained popularity when a plumber by the name of Thomas Crapper patented an improved flushing mechanism in the mid-1800s. (Legend has it that his branded toilets caught the attention of American GIs in World War I, who used his name as a synonym for the devices. The etymology of the word "crap," however, actually traces back to Middle English, with Dutch and Latin roots.) With growing use of flush toilets, the volume of waste substantially increased, exacerbating an already ghastly problem. The Thames in London became so filthy that commercial fishing became unmanageable; once-abundant salmon disappeared entirely by 1833, only repopulating again in the late 20th century after a massive effort.

Rapid modernization attendant with the Industrial Revolution boosted the spread of cholera from its ancient origins near the Ganges River. In 1817, the first pandemic struck India, China, Japan, Southeast Asia, the Middle East, and the East African Coast. With increasing globalization, the disease dispersed along trade routes. A second pandemic swept across Europe and subsequently North Africa and the eastern seaboard of North America. Cholera reached London in 1832, with subsequent major outbreaks in 1841, 1854, and 1866. The prevailing theory of disease held that miasma, a noxious form of "bad air" emanating from rotting matter, caused cholera and other illnesses. Miasmatic theory had predominated since ancient times in Europe, India, and China. An English physician by the name of John Snow had an alternate theory. He believed water, rather than odors, carried these diseases, and he meticulously tracked cases during the 1854 outbreak in Soho, ultimately connecting disease to polluted water sources. Still, the government was unwilling to act on his groundbreaking work, that is, until the unusually hot summer of 1858. High temperatures in July and August brewed the sewage-laden Thames into a putrid soup that emanated such a stench it garnered the moniker the "Great Stink." With the Palace of Westminster sitting on the banks of the Thames, the MPs had first-hand experience with this disagreeable bouquet every day. Acting on a perceived miasmatic threat, this finally spurred them to enact funding to clean up the mess. London's Metropolitan Board of Works went quickly to work constructing a model urban water supply and sanitation system. Waste was rerouted through cement sewers further downstream (a practice that later extended to use barges to ship waste to the ocean). They installed filtration plants to purify incoming water, eventually instituting chlorination for disinfection. When cholera stuck again in 1866, only those communities not yet connected to this new system were affected. This striking observation convinced many that Snow's theory was in fact correct, a shift that was further solidified in 1883 when German physician Robert Koch, the founder of modern bacteriology, discovered *Vibrio cholerae*, the bacteria that cause cholera. (He also identified the anthrax and tuberculosis causative agents.) The Sanitary Revolution was on.

On the other side of the Atlantic, the U.S. was also facing sanitation challenges. In Chicago in the 1840s, the streets were rancid. Sitting at the same level as Lake Michigan, water was not able to drain out of the city. Waste collected to form a ghoulish broth that threatened to hinder the ambitious city's aspirations, contributing to a cholera epidemic in 1849. Chicago dwellers demanded that the City Council rid the city of the filth. The Board of Sewerage Commissioners was created in 1855, and a man named Ellis Chesbrough was appointed its lead engineer. Chesbrough aimed to implement a comprehensive sewer system, which would represent the first of its kind for the country. He faced a major challenge in that the sewer pipes could not be placed under the ground because of the local hydrogeology. Rather than abandon the plan, he undertook what must have seemed a foolish strategy: literally lifting the city higher to build the sewer system under the buildings, which is exactly what they did by jacking up each building by as much as 10 feet. Chesbrough's sewer system included innovations such as manhole covers, which simplified access to clean and repair the pipes. Despite this achievement, the problem was not truly solved. Waste from the sewers and from the city's growing industrial activities flowed into the Chicago River, which emptied into Lake Michigan — the source of drinking water for the city. With polluted water came diseases like typhoid fever, taking a heavy toll. Finally, in 1900, the Sanitary District of Chicago completed the almost inconceivable task of reversing the flow of the Main Stem and South Branch of the river using a series of canal locks, emptying it instead into the newly completed Chicago Sanitary and Ship Canal. As an aside, it is interesting to note that Chicago actually exists because of water. The Chicago River originally drained into Lake Michigan of the Great Lakes. The Great Lakes spill over Niagara Falls and eventually into the Atlantic Ocean. In the first quarter of the 19th century, the U.S. was building the Erie Canal from the Hudson River to Lake Erie. With the locks around the Falls, Chicago connects by water to both the Atlantic and New York Harbor. To the west, the Des Plaines River drains into the Illinois River and eventually into the Mississippi River. This connection gives a water pathway to much of the Midwest, New Orleans, and the Gulf of Mexico. The interesting part about Chicago is that the Des Plaines River and the Chicago River flow within a few miles of each other in a flat, marshy area called the Chicago Portage. This is the closest point connecting the two great watersheds that together encompass virtually all of the land east of the Rocky Mountains. Long before the Sanitary and Ship Canal was constructed, indigenous peoples carried canoes across Chicago Portage to connect between the watersheds. Native Americans revealed this connection to the early European explorers who were searching for trade routes. The earliest known report about Jean Baptiste Point du Sable, Chicago's first non-indigenous permanent resident, comes from an explorer looking for the portage around 1790. OK, enough about our kind of town. Back to sanitation.

The Sanitary Revolution gained further steam with the advent of vaccines, antibiotics, and improved environmental conditions. Human longevity experienced a leap unlike any before in history. Infant mortality, a scourge for millennia, plummeted. The transition to urban living initiated in the Industrial Revolution became sustainable, ushering in the opportunity for an upstart nation to ascend to become a global power.

For the United States to rise to a position of primacy on the world stage, it would have to conquer disparate hydrological environments ranging from the rainy, temperate east to the drier Great Plains, to the drier still arid west and to unify them into a coherent nation. Cutting the country essentially in half is the massive Missouri–Mississippi River system. Its fertile valley is twice the extent of the Nile and Ganges valleys and 20% larger than the Yellow River valley, offering a natural inland transportation path and vast agricultural resources. The high plains and prairies to the west of the Mississippi underwent a farming boom, but unpredictable rains led to yo-yoing population and farm production with cycles of wet years and droughts. As with earlier civilizations, the ability to effectively move food from arable regions to population centers was of crucial importance for development — and canals have proven to be powerful means of such transportation. In the U.S., the Erie Canal slashed the cost of transporting freight, enabling crops from the Midwest to reach not only the eastern seaboard, but also from there to Europe. Settlers poured to the west to take advantage of the newfound demand for American food. Tolls paid by Erie Canal users paid off the considerable construction costs in just about a decade, proving the value of the investment and spurring a canal-building boom — and, soon thereafter, a steam railroad-building boom — across the country. Another, grander canal, far to the south was also transformative for the United States. The French were the first to attempt a connection between the Atlantic and Pacific Oceans across the Isthmus of Panama, but their efforts proved to be unsuccessful. The U.S. took over the project in the early 20th century, completing it in about 10 years. It gave America's navy easy access to both oceans, enabling it to wield power globally and propelling its role in global commerce.

While England's industrial development relied heavily on steam power, the U.S. continued to depend primarily on water power, endlessly improving waterwheel designs. Innovations came fast and furious. American inventor, engineer, and businessman Oliver Evans created a fully automatic, water-powered flour mill — history's first fully automatic industrial process. Eli Whitney's cotton gins were typically water-powered, as was Johann Sutter's saw mill, installed in California (where workers stumbled across gold in the process stream, launching the Gold Rush). Francis Cabot Lowell, using insights from an extended visit to England, created America's first integrated cotton factory, powered of course by waterwheels. His business was so influential the town founded around the factory was even named after him (Lowell, MA). In an effort to draw even more power from the river, engineers at the factory introduced a new evolution in water power technology: the water turbine. These underwater devices channel water through an enclosed passage to turn rotary blades (Figure 4.11), harnessing more energy than earlier technologies. In 1848, James B. Francis, an engineer working in Lowell, MA near the textile factory, improved on these designs to create more efficient "Francis" turbines, still the most common design used today.

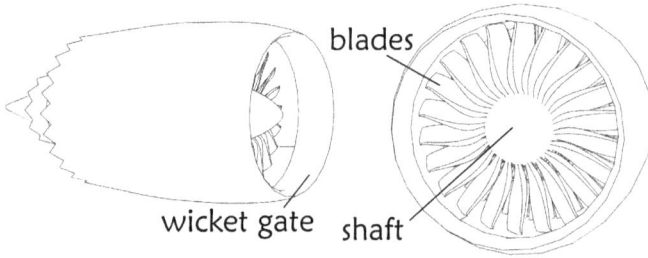

Figure 4.11.   Underwater turbine.

In a world-changing development, as turbine designs became increasingly efficient, they were eventually paired with dynamos to generate electricity. Hydroelectricity demands rapid, voluminous water flow and, not surprisingly, Niagara Falls represented the location for the first large-scale implementation around the dawn of the 20th century. With plentiful electricity, society was transformed. Lighting, radios, aluminum extraction, and refrigeration proliferated, each themselves spurring a collection of other new technologies. Mother Nature, however, does not offer too many places like Niagara Falls. Further expansion of hydroelectricity would have to wait for another breakthrough innovation, in this case spurred by America's water challenges in the Far West.

In the arid west, runoff from snowmelt in the spring is poorly timed for the farming season. There are scant few major rivers, with the Colorado, Columbia, and San Joaquin–Sacramento Rivers representing the bulk, and these are generally far from most arable land. Long before European settlers came to North America, the Hohokam civilization in what is now Arizona had constructed irrigation canals to address this challenge. The Hohokam used water from the Salt and Gila Rivers passed through a canal and weir system. From the 9th to the 15th centuries (when the Hohokam culture collapsed, likely as a result of climatic changes), they maintained widespread irrigation networks rivaling the complexity of those in the ancient Near East, Egypt, and China. Modern western irrigation emerged several hundred years later when the Mormons populated Utah in the mid-1800s. Still, though, the overall water supply was simply too small to reliably sustain larger agricultural production — or aspiring cities on the rise. At the dawn of the 20th century, Los Angeles had outgrown its water supply and was desperate for a solution. City leaders, led by Mayor Fred Eaton and William Mulholland, head of the Bureau of Water Works and Supply, pulled off a notorious water grab (the subject of the 1974 film *Chinatown*), gaining control of the Owens River and building an aqueduct to bring its precious flow to Los Angeles.

A classic strategy to overcome poorly timed seasonal water supply is to construct dams to store water in reservoirs and mete it out as needed throughout the year. Private dam-building efforts sprouted around the west, but they were inescapably small in scale, and they suffered a crippling setback in the spring of 1889. On the other side of the country, upstream of the town of Johnstown, PA, the privately managed South Fork Dam on the Little Conemaugh River experienced a catastrophic failure after several days of heavy rainfall. Eighteen million cubic meters of water rushed out of the Lake Conemaugh reservoir and down the valley into Johnstown with a flow rate that temporarily equaled that of the Mississippi River, killing more than 2,000 people and causing damage of almost half a billion inflation-adjusted dollars. If larger, safer dams were to be built, it was going to require government action.

In his first congressional address, newly sworn-in President Teddy Roosevelt professed his intent to open the American west using federal water conservation and irrigation programs. Soon thereafter, congress passed the 1902 Reclamation Act, providing funding to create the Reclamation Service (later renamed the Bureau of Reclamation), history's biggest water technocracy. Three decades later, Teddy Roosevelt's fifth cousin, President Franklin D. Roosevelt, dedicated a truly massive concrete arch-gravity dam originally known as Boulder Dam in the Black Canyon of the Colorado River. Hoover Dam, as it is now known, dwarfed all previous dams at 726 feet in height, and it created Lake Mead, which can store two times the annual flow of the Colorado. Hoover Dam provided, for the first time, the means to bring a forceful river under almost total control, allocating flow as needed. Importantly, Hoover Dam also provided massive amounts of hydroelectricity. As a multipurpose dam, sales of this electricity subsidized the agricultural irrigation activity, which became a model for future large-scale dam projects. A global frenzy in dam building ensued, with the U.S. leading the way. America's ability to transform river water wealth into economic and military output catapulted the country to superpower status following the Second World War.

Despite gains drawn from giant dams, the country had an insatiable thirst for more water to feed its expanding population and industrial production. In addition to its river systems, the U.S. is fortunate to have water wealth under the ground as well. Sitting under the Great Plains was the colossal Ogallala Aquifer, awaiting a technology capable of pumping its contents to the surface for productive use. Diesel-powered centrifugal pumps provided just that in the middle of the 20th century. Centrifugal pumps transport fluids by the conversion of rotational kinetic energy to hydrodynamic energy of the fluid flow. The fluid enters the pump along the rotating axis and is accelerated by the impeller, flowing radially outward into a discharge pipe (Figure 4.12).

With newfound water abundance, farmers needed an efficient means of spreading the water brought up from a well onto their crops. Center-pivot irrigation achieved this with long chains of sprinkler assemblies

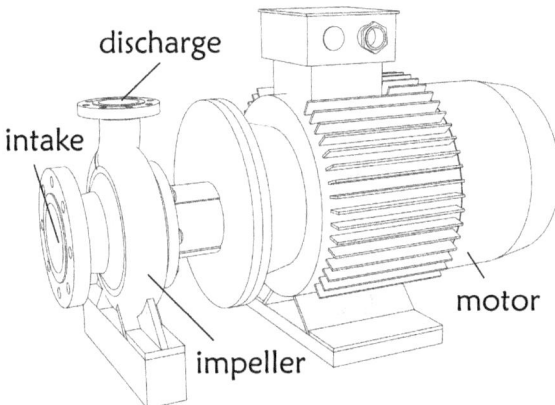

Figure 4.12.   Centrifugal pump.

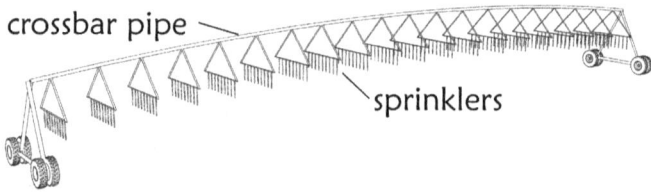

Figure 4.13.   Center-pivot irrigation system.

(Figure 4.13) that could rotate around an extraction well, delivering water from a single source over a wide, circular field. Collectively, these irrigation innovations helped launch the Green Revolution around the world. Food production and, as always, population soared.

As the 20th century drew to a close, it became clear not only in the U.S., but also around the world that the age of water abundance was coming to an end. Nearly every feasibly dammable river on Earth had been dammed. Many great rivers were so heavily exploited that they no longer reached the sea; many other surface water sources were being polluted, rendering them useless for irrigation or other human use. Damming has several potential negative consequences including the displacement of communities, disruption of migration of aquatic species, perturbation of sedimentation along the river bottom, and shifts in the natural borders of the river. Dams, *per se*, do not decrease aggregate water flow, but rather divert water from the river. Two notable examples are the Colorado River in the southwestern United States where the river no longer reaches the Gulf of California and the Aral Sea in southwest Asia, where water diversion from its tributary rivers has caused this inland freshwater sea to shrink to a small fraction of its original size. As surface resources became increasingly depleted, groundwater mining expanded far beyond nature's ability to replenish the aquifers. Human consumption had expanded beyond the planet's supply of readily accessible clean water. Many of the most arid regions of the world are also heavily populated and poor, with little hope for feeding their people in the future. As we entered the 21st century, it became clear that the golden age of water was over. A new era of freshwater scarcity materialized, representing perhaps the greatest global risk we face this century. In the next section, we will cover the myriad ways in which we use water as a society, highlighting the direness of the situation in which we find ourselves. Later in the book, however, we will lay out paths forward to a sustainable water future.

The many uses of water

In the age of scarcity, taking a look at where and how we use water is a necessary step toward holistic, sustainable management of this precious resource. From 1900 to 2000, global population tripled while water use increased a whopping six-fold. Clearly, direct human consumption in the home cannot explain such a discrepancy. The truth is that domestic uses of water, although the most visible to us, represent only a small fraction of our overall water use as a society. Many factors, including demographics, economic and regulatory trends, climate changes, and new technologies can influence water use. Here we need to make an important distinction about water — it has flavors, and we are not referring to lemon or strawberry. Fresh water is classified as being "blue" or "green." Blue water is the surface water in lakes and rivers as well as the water under the ground in aquifers. This is the primary source for human use. Green water is precipitation that naturally falls on the soil and then evaporates or becomes incorporated within plants. This is the primary source for natural ecosystems as well as for rainfed agriculture. Another important distinction about water is the difference between water use (or "withdrawal") and water consumption. Much of the water that is withdrawn from blue sources is returned to rivers, lakes, and aquifers, although typically with significantly reduced quality resulting from its use. (This polluted water has its own color classification: gray.) This water is not considered to be consumed.

You can compare the overall global availability of renewable, blue water resources, and you may be surprised to see that it is *far larger* than our consumption for the foreseeable future. So what's the deal with this age-of-scarcity nonsense? Unfortunately, most of the global water resources are in geographies well separated from where the people are, such as in Canada or the Amazon. Also, as we described in the previous section, annual precipitation is often not timed well with use patterns, such as with the seasonal monsoons in southeast Asia. In keeping with their tradition for mammoth hydraulic engineering projects, the Chinese have recently attempted to reshape Mother Nature's designs to overcome such resource and population mismatch. In the 1950s, Mao Zedong returned to the challenge of a water-rich south and water-poor north that has plagued the region for millennia. The ancient Chinese constructed the Grand Canal to move goods between these regions on a waterway; Mao proposed the far more ambitious concept of actually moving water. It was not until the 2000s, however, that The South–North Water Transfer Project finally began. This ongoing mega-project aspires to transport more than 40 billion cubic meters of fresh water from the Yangtze River to the arid north each year, using three canal systems — including one following the course of the Grand Canal itself. The project has come under intense criticism for its immense cost and potential environmental and demographic impacts, and only time will tell if the ambitious plan pays off. It is unlikely that any other nation will undertake such a massive water transport project again. Alternative solutions to aligning water resources and water demands are desperately needed.

Another important consideration is that we humans are not the only organisms on the planet that need blue water. Natural ecosystems also depend upon water for their wellbeing — and we depend, in turn, upon them. With all these reflections in mind, let's examine the various ways in which we use water. When numbers are useful, we will use the United States as a representative developed country since the U.S. Geological Survey has maintained detailed data records for more than 50 years. Even within a given country, water use can vary dramatically, but having an example set of data such as this offers a useful tool for unveiling important, and often universal, issues surrounding water.

Figure 4.14 shows the breakdown of total water use in the U.S. The lion's share of use is for generation of electricity in power plants, representing nearly half of all water withdrawals. However, most of this

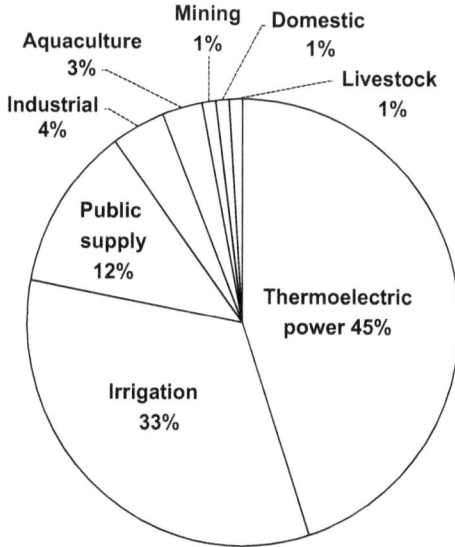

Figure 4.14.   Total water use in the United States (2010).

water is returned to waterways after flowing through the power plant, with only a small fraction consumed in the form of steam that turns turbines and other evaporative losses. The largest overall *consumptive* use of water is for irrigation. Illustrating the geographic complexity of water use, more than half of all water withdrawals in the United States occur in only 12 of the states (CA, TX, ID, FL, IL, NC, AK, CO, MI, NY, AL, and OH), with California alone representing 11% of the total. Most of California's withdrawal goes to irrigate farms, whereas power production is dominant in many of the other water-greedy states.

It is also informative to observe how water use has evolved over time. Figure 4.15 displays the trend in surface and groundwater use in the U.S. from 1950 through 2010, with the nation's population trend superimposed. Population has risen steadily over this period, which translates into greater domestic demand for food, industrial production, and other water-reliant sectors. From 1950–1980, water withdrawals mirrored population growth in keeping with this correlation. Interestingly though, in the following decades, water withdrawals have been approximately

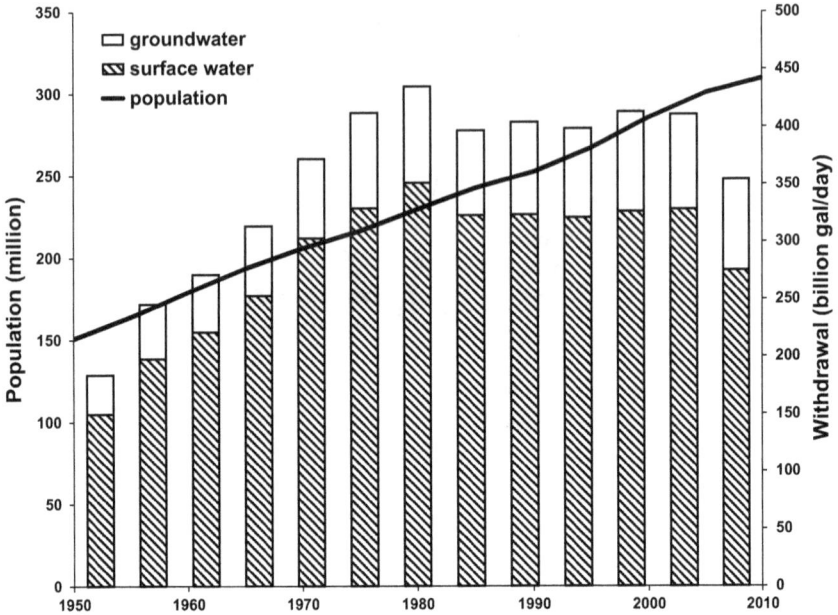

Figure 4.15.   Trends in population and freshwater withdrawal in the United States.

constant. The principal explanation for such a disparity is a combination of efficiency gains and water conservation efforts. Advances in farm and irrigations practices and in crop seed drought resistance have been major contributors to water efficiency. The ratio of surface and groundwater withdrawals has remained roughly flat, with ~20% coming from groundwater throughout this period. Not all water use sectors exploit the same ratio, however. Virtually all self-supplied homes rely on groundwater; aquifers also represent a major source for irrigation and livestock. There is as well a geographic trend of drier states, such as those in the American west, relying more heavily on groundwater for their needs. In the remainder of this section, we will briefly explore each of the sectors of water use in detail.

Electricity production is intimately tied to water, so much so that the community often refers to an "energy-water nexus." One pillar of the nexus is the use of water in steam-driven turbines operating by electromagnetic induction. Heat comes from combustion of fossil fuels, nuclear fission, geothermal sources, biomass, or, increasingly, concentrated sunlight. This heat boils water to create steam, which is forced through turbines. Electromagnetic induction involves the use of a dynamo or alternator to transform turbine rotational kinetic energy into electricity. This is the most common means by which electricity is generated around the world. There is a second way in which water and turbines can generate electricity, which was described in the previous section and involves using a dam on a river to create artificial reservoirs. Water flows through underwater turbines, turning gravitational energy into rotational kinetic energy. Such hydroelectricity is a popular, low-carbon energy source, and its use has been increasing globally for many years (Figure 4.16). Today hydropower represents the most widely used renewable energy source, with large footprints in countries like China, Canada, Brazil, and the United States. Surpassing the Hoover Dam,

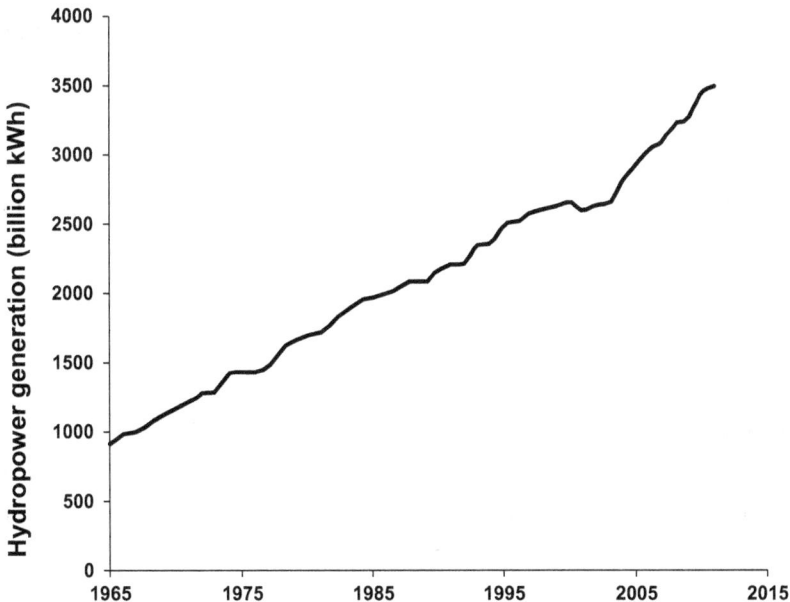

Figure 4.16.   Global hydroelectricity generation from 1965–2011.

the pioneering giant power station, are the likes of Itaipu Dam on the Paraná River in South America and the massive Three Gorges Dam on the Yangtze. There is growing concern, however, over the proliferation of hydroelectric dams. Creation of new facilities requires massive investment cost, inundates habitats or archeological sites, and often displaces local populations with the formation of large reservoirs. Large hydroelectric dams also can obstruct the migration of fish, affecting populations indirectly, and entrain fish into the power generation systems, directly killing them.

We now turn our attention to the largest consumptive use of water: irrigation. Unlike in many other uses of water, much of the water used in this sector is lost by evaporation into the air and evapotranspiration from plants (or, unfortunately, by leaks). Without irrigation, feeding a global population of 7.5 billion and growing would be inconceivable. Not surprisingly, countries with the largest populations are often those using the most water to grow food. China, India, and the U.S., for example, represent the three most populous nations and the three largest water users for irrigation. Each nation relies on a specific set of water resources and each also uses a unique combination of irrigation methods for applying the water to crops. Depending on the local resources, a nation will draw water for irrigation from various sources in different proportions. India's irrigation water comes mostly from aquifers, whereas Indonesia is almost exclusively drawing on surface water sources (Figure 4.17).

In broad strokes, irrigation methods can be classified as surface, sprinkler, localized, or sub-irrigation. In a surface irrigation system, which is by far the most widespread method of irrigation (Figure 4.18), water is transported by gravity across an agricultural plot and infiltrated into the soil. A common implementation of surface irrigation is a level basin system in which water is applied quickly to a large plot of land, flooding it and allowing it to slowly infiltrate. Furrows represent a second class of surface irrigation. Furrow irrigation is performed by creating parallel channels along the field length following the natural slope of the land.

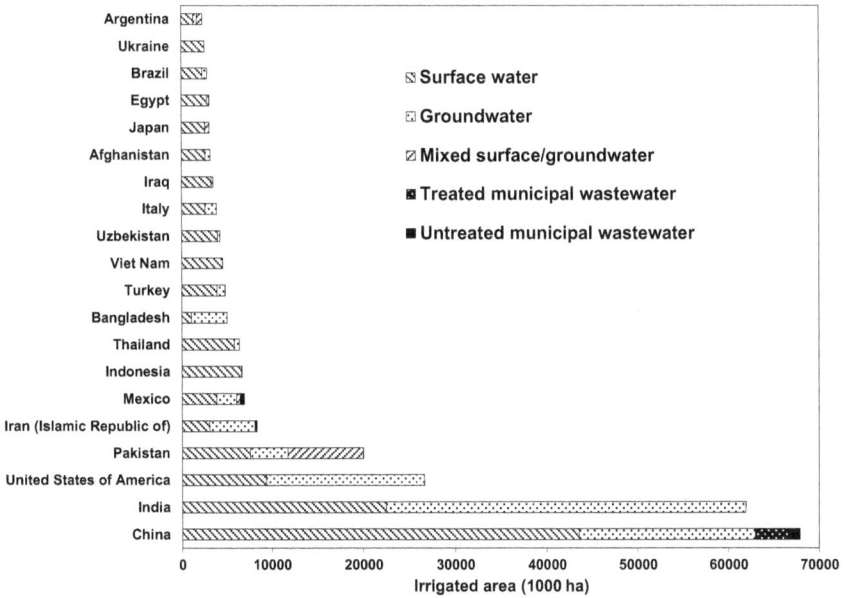

Figure 4.17. Water sources for irrigated farmland in the top 20 consuming countries. (Data from AQUASTAT Main Database — Food and Agriculture Organization of the United Nations, 2016.)

Water is introduced to the top end of each furrow, from where it flows down the field under the influence of gravity and is distributed via ditches, tubes, or other means. Numerous factors influence the rate of water movement such as slope, surface roughness, furrow shape, inflow rate, and properties of the soil itself. A third subdivision of surface irrigation is borderstrip, also known as border check or bay irrigation. Borderstrip is a hybrid of level basin and furrow irrigation. The field is divided into a series of "bays" or strips with each bay separated by raised earth check banks. Field water efficiency of surface irrigation is usually lower than other forms of irrigation, with optimal efficiencies in the range of 80%, but more typically around 60%.

In sprinkler irrigation, water is routed to one or more locations within an agricultural field and distributed from those locations by high-pressure overhead sprinklers or guns. A common implementation of sprinkler irrigation is the center pivot system discussed earlier in this chapter. In

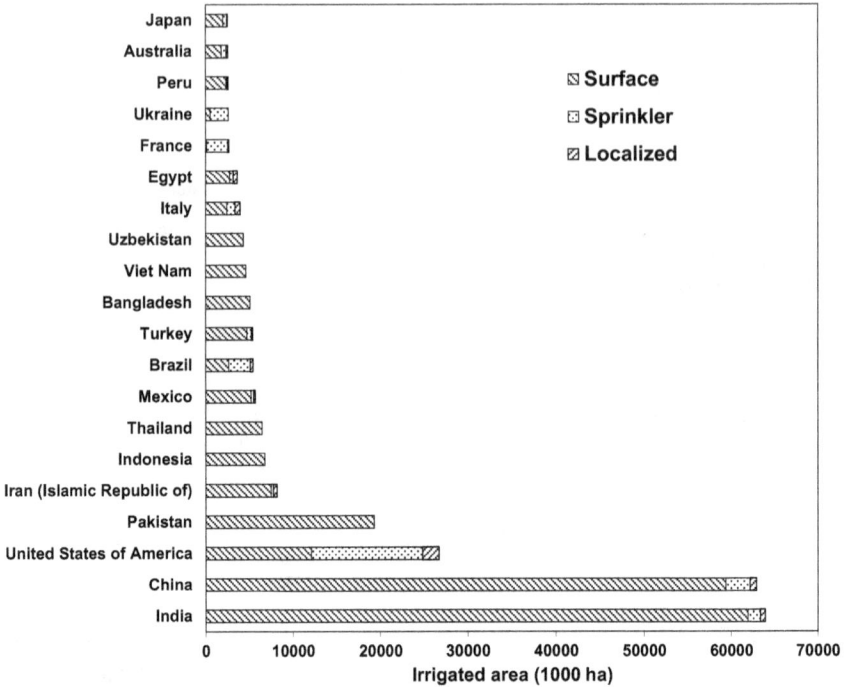

Figure 4.18.   Application technology for irrigated farmland in the top 20 consuming countries. (Data from AQUASTAT Main Database — Food and Agriculture Organization of the United Nations, 2016.)

such systems, several segments of pipe are joined together and mounted on wheeled towers with sprinklers positioned along its length. The system rotates with the water feeding in from the central point, so such fields typically have crops planted in circular patterns. Alternatively, one can use a lateral-move sprinkler system in which a series of pipes are joined, each with a wheel fastened to its midpoint and sprinklers along its length. Water enters through a hose at one end. After adequate irrigation has been applied to a particular strip of the field, the hose is removed and the assembly shifted to a different position. The process is repeated in a pattern to irrigate an entire field. Lateral-move systems are less expensive to install than a center pivot, but substantially more labor-intensive to operate. They are therefore generally used for small or oddly shaped fields or in regions with inexpensive labor. Field application efficiency for sprinkler systems are generally around 75%.

In localized irrigation, water passes at low pressure through a network of pipes and is applied in small quantities directly to each plant. A popular realization of a localized method is drip irrigation, sometimes called micro or trickle irrigation, where water is delivered one drop at a time to a plant's root zone. Drip irrigation is the most water-efficient option, with field water efficiencies often approaching 90%. Drip irrigation is often combined with plastic mulch, reducing evaporation losses while also providing a means of delivering fertilizer — this is known as "fertigation." The final category, sub-irrigation, is only applicable in regions with a high water table, usually in a river valley. In this technique, a farmer artificially raises the water table to access the root zone of the crops. A system of pumps, canals, weirs, and gates enables control over the water level in a network of ditches that secondarily influence the local water table. This method is also used in commercial greenhouses.

The third largest fraction of water withdrawals (12%) is dedicated to public supply, perhaps more familiar to most as city or county water departments, although these entities can also be privately owned. There are more than 150,000 public water systems in the U.S. alone and they range in size from serving a few dozen homes up to 10 million people. These facilities withdraw water from rivers, lakes, seas (requiring desalination), or aquifers, treat it, and deliver it to homes and businesses through networks of water mains. As is so often the case with water, the precise contributions to the public supply from these different sources vary considerably around the world — often even within one country. Loosely speaking, Europe relies most heavily on groundwater (~75%), followed by North America (~65%), Asia-Pacific (~30%), Latin America, and Australia (~15%). Within the U.S., where roughly 90% of the population rely on public water supply (as opposed to a private well), most states depend predominantly on surface water, but a few populous states such as Florida, California, and Texas are groundwater-dominant, skewing the national average. After withdrawing water, in most cases it must be treated to bring it up to drinking water standards. Water treatment typically involves purification, disinfection, and sometimes fluoridation. Treated water then either flows by gravity or is pumped to reservoirs, which can be elevated, as in the case of water towers, or on the ground. Globally, approximately half of the population has access to a piped water supply, but in many regions of the world, the quality of service is dreadfully insufficient. Water supply should exhibit continuity, high water quality, and reliable pressure. Responsibility for ensuring service falls on regulatory agencies as well as service providers. In the United States, the Environmental Protection Agency (EPA) is responsible for water and sanitation policy and standard setting, although individual states and regions frequently impose stricter regulations than nationally mandated. In other countries, responsibility for water policy is assigned to one or more of agencies such as the Ministry of Environment, Ministry of Health, Ministry of Public Works, Ministry of Economy, or Ministry of Energy. Several countries, such as India, Bolivia, and Jordan, even have a Ministry of Water.

Within the home, water finds many uses, and the relative amounts of water directed to particular uses can vary dramatically depending on the location. For example, outdoor use for irrigation of landscaping consumes a larger share in drier climates where natural precipitation is insufficient to maintain much greenery. Figure 4.19 depicts a typical breakdown of how water is used in a home. Having such data in hand assists in identifying places where efficiency could have the biggest impact on overall water consumption. Innovations such as dual-flush toilets and low-flow faucets are clearly important, but perhaps not so apparent is that shifting to native plants that require little or no watering for the landscaping around one's home could dwarf the water savings from in-the-home technologies.

A separate category of withdrawals is those homes that are self-supplied, accounting for 1% of overall withdrawals. Nearly all self-supplied domestic users access fresh groundwater sources using wells, which can be dug, driven, or drilled. Shallow wells (dug or driven) can often supply drinking water at low cost. However, impurities from the surface

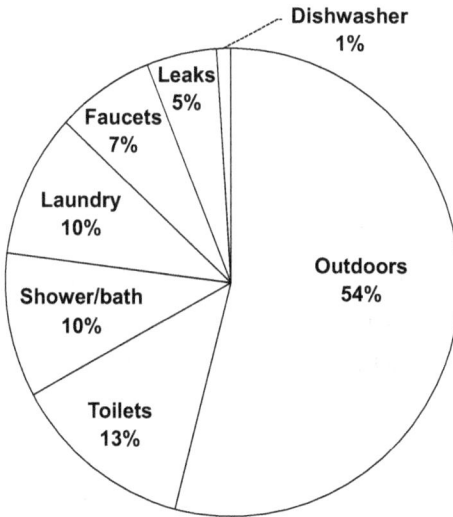

Figure 4.19.   Typical breakdown of domestic water use.

(pathogens or chemicals) can readily reach shallow sources, which increases risk of contamination compared to deeper wells.

Marching further down the list of heavy water withdrawals, next comes industry at 4% of the total. Every single manufactured product requires water during at least some part of its production process. Water can be used for fabricating, washing, cooling, transporting, and even incorporation directly into the product itself or for sanitation at the manufacturing facility. Food processing, papermaking, chemical production, petroleum refining, and metal manufacturing represent some of the most water-intensive industries. Not surprisingly, then, regions with a heavy presence of these industries often consume disproportionately large shares of water for industrial use. In the U.S., for example, Louisiana, Indiana, and Texas account for 40% of the entire nation's industrial water withdrawals. Of course, some industries obtain their water from the same public supply as a typical homeowner, but most are actually self-supplied both to limit costs and to tailor the water quality to their specific needs rather than relying on standard drinking water regulations. A fascinating illustration of this is the semiconductor industry. Semiconductors are the materials that serve as the foundation of modern electronics; they are found in microchips, which provide the processing power and memory, as well as in solar cells and countless other high-tech components. Water is essential to the manufacture of semiconductors. Silicon wafers are transformed into integrated circuits over a tedious series of steps, many of which require rinsing with water before moving to the next step. Because even one microscopic fleck of impurity could ruin such a wafer, most of the rinsing water must be ultrapure water (UPW), which is thousands of times purer than drinking water. Indeed, UPW is so distinct from municipal water that it is considered an industrial solvent. UPW systems are expensive to build and operate, representing a measurable fraction of the capital and operating costs of these pricey facilities. A single wafer requires over 2,000 gallons of water, and a typical large fabrication facility will process tens of thousands of wafers each month, representing an overall water use comparable to a small city.

Similar to agriculture, some industries are such intensive water users that they are considered a separate category when evaluating water withdrawals. Another such example is aquaculture, representing about 3% of overall withdrawals. Farming of fish, crustaceans, seaweed, and other aquatic organisms involves cultivating populations of these species under controlled conditions, as opposed to commercial fishing operations in which wild creatures are caught. Statistics provided by the Fisheries Department of the Food and Agriculture Organization of the United Nations (FAO) reveal total aquaculture production of over 100 billion pounds per year. Global output from aquaculture operations is reported to represent about half of the fish and shellfish consumed by humans, although it should be noted that these statistics are self-reported by countries and the numbers are often under dispute. In the context of expanding global population and a projected decline of output from capture fisheries due to climate change, pollution, and overfishing, aquaculture production will need to increase dramatically in the coming decades. Fortunately, much of the water use in aquaculture is in flowthrough raceway operations that return the water to the environment, much like water withdrawals used in power generation described earlier in this section. Moreover, many operators are instituting water efficiency practices, including water treatment and recycling. There are other reasons for concern about aquaculture sustainability, such as the fact that products from several pounds of wild fish are used to produce one pound of a piscivorous fish like salmon, but water sustainability may be achievable in this industry.

Another major water consumer is the livestock industry. Here there is an important distinction to make, into which we will take a deeper dive in the next section: direct water use vs. so-called virtual water use. Livestock operations represent a bit less than 1% of overall water withdrawals for direct use. Water use in this industry includes watering of the animals, cooling of facilities for the animals and animal products such as milk, dairy sanitation and wash down activities, and animal waste disposal. The majority of withdrawals come from groundwater sources, and the activity is heavily localized in certain regions. In the U.S., Texas, California, Oklahoma, and North Carolina together account for 35% of livestock water withdrawals. These direct uses represent a lot of water, but there is actually a much larger, hidden use of water in livestock operations. This hidden water is in the food that livestock eat. An animal's efficiency to turn food it eats into body mass is called its feed conversion ratio (FCR). FCRs vary depending on the type of animal, roughly 7:1 for beef, 5:1 for pork, and 2.5:1 for poultry. The larger the animal, the greater percentage of the animal's body mass is inedible material like bone or skin. Animals such as beef cows represent massive amounts of water because a typical beef cow eats thousands of pounds of corn, soybeans, and hay during its lifetime. As we discussed earlier, those crops consume large amounts of water themselves, and there is a colossal efficiency loss in turning those crops into meat.

Finally, the last industry to be singled out is mining, accounting for 1% of total water withdrawals. Water is essential to mineral processing. Minerals may be solid (coal, iron, gravel), liquids (oil), or gas (methane), and these commodities are critical to vast industries ranging from pharmaceuticals to high-tech products. Water is needed for quarrying, controlling dust, milling (includes washing and flotation), injection for secondary oil recovery, and other operations. The amount of water required varies depending on a mine's size, the mineral being extracted, and the extraction process. Metal mines that chemically process ore to concentrate metals such as gold or copper use far more water than non-metal operations such as salt or gravel mines. Moreover, as ore grades decline the amount of water required increases because more intensive processing is necessary. Water is not always readily available for mining operations. In many locations where mining operations occur, such as in remote Australia or northern Chile, water is a scarce resource. In many cases, groundwater is tapped for this purpose, and these sources are often saline, meaning additional treatment may be required prior to use in the operation. Again, withdrawals are often concentrated in specific geographic locations. Texas, Minnesota, and California account for 34% of mining water withdrawals within the U.S. Mining has an additional challenge with water. Depending on the operation, some injected water returns to the surface as flowback water. Depending on the mining product and the subsurface chemistry, flowback water can contain large amounts of heavy metals or be strongly acidic. Frequently, post-extraction treatment of mining water can be very expensive and requires deep reinjection.

The direct uses of water we have laid out in this section result in the production of numerous products that we all use every day, meaning that each product has a water footprint. In the following section we will take a closer look at this virtual water and how it impacts commerce on a global scale.

Virtual water

Every single product manufactured by humans has a water footprint — a measure of the amount of water consumed and polluted over all the processing steps of its production. This concept goes by many names: virtual water, embedded water, and embodied water; we will stick with "virtual water" in most cases for consistency. Before a t-shirt can be manufactured, cotton must be grown, ginned, and spun into fibers that are cleaned, woven, dyed, and sewn to produce the final product. Transportation of goods is also important to consider. Each of these stages carries with it a direct water footprint, broken down by blue, green, and gray sources, which gets folded into the indirect water footprint of all the subsequent stages. When such an analysis is performed carefully and comprehensively, the final tally offers an appraisal of the efficiency and sustainability of a given product. Blue and green water footprints point to the amount of water resource consumed, and gray water footprints indicate the burden put on water quality. Such information is vital in determining how best to utilize limited global resources and what practices are likely to be sustainable.

As helpful as the virtual water concept is, there are nuances and limitations to consider. It is crucial to remember that the time and location of water use can matter as much as the absolute amounts of water involved. A water-rich region or use during the wet season, for example, can offer more sustainable supplies. A further essential clarification is whether the blue water footprint comes from surface or groundwater sources, with the former typically being more sustainable than the latter. Virtual water also implicitly assumes water saved by not using it for one product would be available for another use, which is not necessarily the case. Water cannot easily be moved over large distances, and different products generally require different water quality in their manufacturing. With these stipulations in mind, let's take a look at the (globally averaged) water footprints of some common food products, since food represents such a large fraction of overall water withdrawals. Figure 4.20 breaks down the virtual water associated with each product into its blue, green, and gray footprints. Even this small selection of products reveals a number of informative insights into virtual water.

Generally speaking, more processing translates to a larger water footprint, so heavily processed products such as cheese and bread involve more water use than basic ingredients like milk or wheat. Even among basic agricultural products, some items consume much greater amounts of water than do others. Typical salad ingredients are relatively water-efficient

(gal/lb)	blue	green	gray
olive	61.4	296.2	7.2
peanut	16.7	296.7	20.0
date	150.1	111.9	10.9
mango	43.1	157.5	15.1
corn	10.2	112.7	23.4
peach	22.9	69.8	16.4
apple	15.8	67.0	14.8
banana	11.4	79.5	3.8
orange	13.4	48.3	6.0
cucumber	5.1	24.5	12.7
potato	3.8	22.7	7.6
lettuce	3.4	15.9	9.1
tomato	7.7	12.8	5.1
beef	73.9	1736.3	55.4
sheep meat	62.4	1172.8	12.5
pork	57.4	588.4	71.8
goat meat	39.7	621.9	0.0
chicken	36.3	425.0	57.0

(gal/8 oz)	blue	green	gray
coffee	0.7	63.4	2.0
milk	5.1	54.2	4.5
wine	8.7	38.2	7.6
beer	1.1	15.7	1.7

(gal/lb)	blue	green	gray
chocolate	20.6	2019.3	20.6
butter	53.2	565.6	46.6
cheese	30.5	323.7	26.7
rice	59.8	203.5	32.9
pasta	42.1	155.1	24.4
sugar (cane)	57.7	140.9	12.8
bread	36.6	134.9	21.2
sugar (beet)	20.9	68.3	20.9
egg	0.1	1.1	0.2

Figure 4.20.   Virtual water footprint of various foods and drinks.

whereas products grown on trees like pitted fruits tend to use more water. Looking in more detail at the makeup of water footprints, it is clear that the ratio of blue to green water varies dramatically between different products. Peanuts are grown almost exclusively with natural precipitation, whereas dates require substantial artificial irrigation. One obvious conclusion from these data is that dairy and, especially, meat involve vastly more virtual water than do cereals, fruits, and vegetables. As discussed earlier in the chapter, this is because the livestock consume large amounts of agricultural products during their life. Among meats, poultry has greater water efficiency than beef. Another important point is that how you make a product can matter a lot. Sugar extracted from cane, for example, has a notably larger water footprint than sugar from beets.

Water resources, both in terms of quantity and quality, vary markedly by location. On a global scale, water scarcity can be addressed, in part, by moving goods from place to place. In the context of escalating water demand, finding ways to share water (and virtual water) equitably and sustainably around the planet is a daunting challenge. Each country can be assigned a water footprint with regard to its production and to its consumption. Products produced within the country count toward its water footprint of production, regardless of where the goods are ultimately consumed. The water footprint of consumption, on the other hand, is the total of goods and services consumed by people living within that country, regardless of the origin of the products. On a per capita basis, water consumption varies considerably around the world as a result of dissimilar gross national income, cultural practices/diets, and even climate-driven evaporation. While highly developed countries such as the U.S. have both large total water use and a large per capita water footprint, poor and developing countries tend to have much larger water footprints per unit of national income. This discrepancy suggests that developing countries do not use water as efficiently as more advanced economies. When viewed collectively, these two perspectives of water footprint — production and consumption — offer powerful understanding of a country's water resources and dependence upon its trading partners. Calculating the fraction of a country's consumptive water footprint that is within its borders is critical when evaluating overall security issues. Figure 4.21 offers a schematic to tell part of such a story. The United States consumes many goods that are produced within its borders, but it also exports large amounts of virtual water, for example, in the form of maize sent to neighboring Mexico, and it also imports virtual water, for example, in the form of coffee from Columbia. Interestingly, the U.S. is both the largest virtual water exporter *and* the largest virtual water importer. Only when the full balance sheet is done can a country establish a complete picture.

Virtual water trade, therefore, represents an important pillar of geopolitics. Figure 4.22 provides a country-level representation of the global virtual water trade. Countries with ample water resources tend to be virtual water exporters and those in arid regions must import virtual water, particularly in the form of food to nourish their domestic population. Alfalfa hay grown in Argentina is shipped to the Middle East to feed dairy cattle.

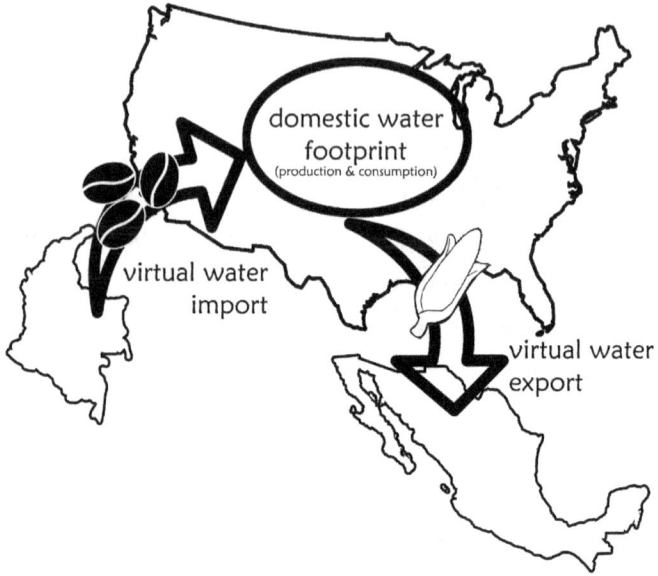

Figure 4.21.   National water footprint.

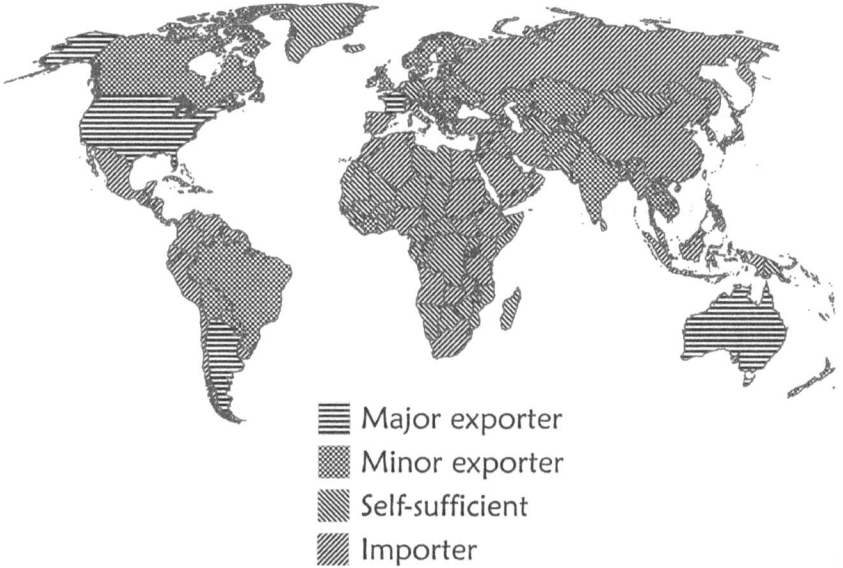

≡ Major exporter
▦ Minor exporter
▨ Self-sufficient
▨ Importer

Figure 4.22.   Global virtual water trade.

China imports soybeans from the U.S. and Brazil. Chile and Peru depend on cereals, nuts, and other goods from trading partners. Without this movement of virtual water, many countries would not exist in their current form.

From a macroeconomic perspective, water-scarce nations can be expected to import water-intensive products and to export water-extensive products, thereby relieving pressure on their domestic resources. Those nations with water wealth will seek to profit by exporting water-intensive goods. Integrated across the globe, international virtual water trade exceeds 400 trillion gallons per year, approximately 60% of which is crops and crop products, with the remainder more or less evenly split between livestock products and industrial products. Some of the countries with the largest net import of virtual water are Japan, Italy, and South Korea; major exporters include Australia, Argentina, and the United States. Digging deeper can reveal interesting insights such as the fact that China has a net import of virtual water for agricultural products but a net export of virtual water for industrial products. A country like the Netherlands is a net virtual water importer, but has a net export in virtual water associated with livestock products.

Despite its significance, the global virtual water trade is almost entirely unregulated. You cannot see containers of water at the border when the water is embedded in the production process. International water dependencies are considerable and are likely to increase in the future. Water resource policies will need to take international and interregional virtual water flows into account. It is important that such policies incorporate food security, natural resource management, sustainability, and the universal human right to a basic standard of living. Policies are likely to originate not only in multilateral and bilateral national forums, but also from multi-stakeholder NGOs and business groups, making jurisdictional issues increasingly complex. Despite the complexity, the current unregulated virtual water trade is surely not providing for the most sustainable and fair use of our global water resources, so clearly the challenge must be tackled. We will explore broader water policy issues in more detail in the next chapter.

05 **Water is Government**

Chapter

Water resource policy

Water pays no attention to our political boundaries, with most resources crossing many borders. Worldwide, there are 276 transboundary river basins and over 200 transboundary aquifers; 148 countries include territory within these transboundary river basins. Therefore, policies will almost always require negotiation among numerous entities — both between and within sovereign nations. Policies surrounding water comprise a broad array of topics, spanning identification, collection, treatment, distribution, allocation, use, disposal, finance, and sustainability. On the international stage, momentum for establishing more coherent water policies gathered steam in the 1970s at the United Nations' Water Conference at Mar del Plata, which designated the 1980s as the International Drinking Water Supply and Sanitation Decade. This declaration was spurred by concern that a large part of the world's population lacked access to ample, safe water supplies and an even greater part lacked adequate sanitation. This first water decade substantially improved access to both drinking water and sanitation for many people in the developing world. However, population growth and urbanization continue unabated, heightening the challenge. Following the first International Drinking Water Supply and Sanitation Decade, leadership has transitioned to the World Water Assessment Programme, a joint program of the UN and its member states, which performs a biennial assessment of the status of water resources around the world. Moreover, global attention on water issues was reinvigorated in the period 2005–2015, which served officially as the second International Decade for Action "Water for Life." A larger fraction of people in low- and middle-income countries have access to a mobile phone than to a safe source of water. Today, there are still over one billion people with inadequate access to water and over two billion without suitable sanitation.

Multilateral and bilateral water agreements have a long history. In 2500 BC, the two Sumerian city-states of Lagash and Umma settled a dispute over Tigris River water with an agreement — purportedly the first treaty *of any kind* in human history. In the many years since that time, thousands of treaties associated with international water resources have been crafted, primarily addressing navigation and boundary demarcation. In recent years, the focus has been shifting away from navigation to issues surrounding use and sustainability. Figure 5.1 provides a breakdown of the subjects for a sample of approximately 150 transboundary water resource agreements (data are from the UN Human Development Report 2006). In the modern era, hydroelectric power, which can provide enormous benefits for a country building a dam, but at the same time can substantially

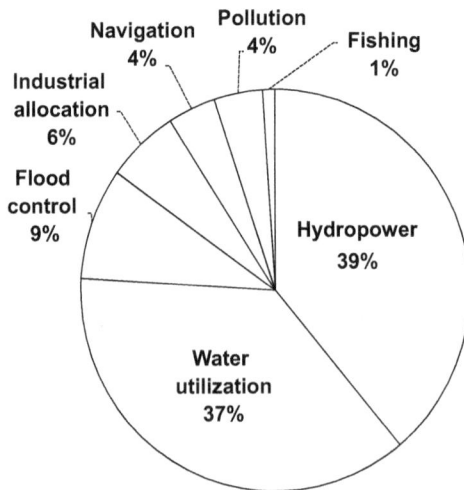

Figure 5.1.   Distribution of subjects for transboundary water resource agreements.

reduce access to water and fish in downstream nations, dominates such agreements. Other policy issues that often arise are flood control, use of freshwater sources, and pollution challenges such as industrial and agricultural runoff or wastewater discharge.

In the middle of the 20th century, even countries experiencing profound political tensions found ways to cooperate through water agreements. Israel and Jordan have long shared Jordan River resources. The countries in Southeast Asia have collaborated within the Mekong River Commission since 1957, including throughout the Vietnam War. Likewise, India and Pakistan have sustained cooperation through two wars via the Indus River Commission, established by a treaty signed in 1960 by Prime Minister of India Jawaharlal Nehru and President of Pakistan Ayub Khan. More recently, Africa has joined the effort. Nine countries comprising the Niger River Basin (Niger, Benin, Chad, Guinea, Côte d'Ivoire, Mali, Nigeria, Cameroon, and Burkina Faso) formed the Niger River Commission in the 1960s, which later transformed into the Niger Basin Authority in the 1980s. The Nile River Basin spans another nine countries (Egypt, Sudan, Ethiopia, Uganda, Kenya, Tanzania, Burundi, Rwanda, and the Democratic Republic of Congo), all of which joined a framework agreement under the Nile Basin Initiative in 1999. Cases such as the Niger and Nile agreements share two common aspects of international cooperation in water resource management: third-party support with sufficient funding and trust, and sustained engagement through a formal institution. That said, despite thousands of treaties being executed, the content and structure of such agreements remain incomplete. International watercourse agreements must establish firm verification, enforcement, and conflict resolution mechanisms. Water allocations and quality standards must on the one hand be equitable and reasonable and, on the other hand, flexible to accommodate changing hydrological and social conditions, particularly in the context of a changing climate.

A rare example of such a progressive agreement, involving both international and domestic aspects, is that governing the Great Lakes

Basin — the largest accessible source of freshwater on the planet. (Note that Lake Baikal in remote southern Siberia is actually the largest freshwater lake in the world by volume, but its geographic isolation renders it practically inaccessible.) The United States and Canada have an extended history of collaborating to address issues facing waters that cross their shared boundary, beginning with the Boundary Waters Treaty of 1909. Momentum for such a treaty sprouted from disagreements apportioning the water from various boundary rivers as well as Chicago's manipulation of basin flows associated with the Chicago River and Lake Michigan (see Chapter 4). The Boundary Waters Treaty was meant to provide a framework to settle such disputes. A landmark extension was signed in 1972 that focused on the Great Lakes, known as the Great Lakes Water Quality Agreement. This document leveraged many extant programs and respected each country's culture of addressing water quality issues. Importantly, it also incorporated adaptability to tackle new challenges as they emerge as well as periodic reviews of the operation and effectiveness of the agreement to proactively identify such challenges. Indeed, there have been numerous revisions of the agreement in the subsequent decades. Exemplifying the growing importance of water overall and of the regional resources specifically, a new agreement was signed in 2005 to incorporate broader issues beyond just water quality. The Great Lakes–Saint Lawrence River Basin Sustainable Water Resources Agreement (that itself builds on the 1985 Great Lakes Charter) is a good-faith agreement among the governors of Illinois, Indiana, Michigan, Minnesota, New York, Ohio, Pennsylvania, and Wisconsin (states bordering the Great Lakes; Figure 5.2) and the premiers of the Canadian provinces of Ontario and Québec. In order to establish a structure for the participating U.S. states to implement their commitments under the agreement, they formed the domestic interstate, legally binding Great Lakes–St. Lawrence River Basin Water Resources Compact. The compact specifies how the states manage use of the water supply in the Great Lakes Basin. In a further sign of cooperation around water in the region, these U.S. states and Canadian provinces formed a working group called the Ten-State Standards. The

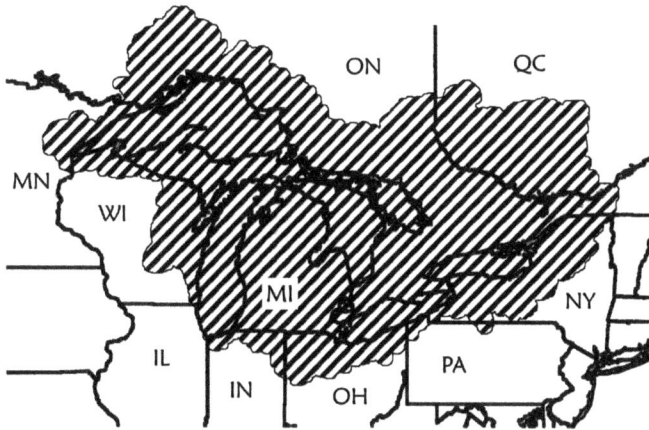

Figure 5.2.   Great Lakes Basin (shaded area) and bordering U.S. states and
Canadian provinces.

Ten-State Standards shares technology and policy recommendations about
wastewater treatment and management.

In addition to international and interstate governmental agreements,
a number of non-governmental organizations (NGOs) weigh in on policies
related to the Great Lakes, in addition to their general advocacy for protection,
cleanup, and education. NGOs such as the Alliance for the Great Lakes and
the Great Lakes and St. Lawrence Cities Initiative (itself led by mayors of
cities in the region) have long worked to implement policies governing the
basin. Many NGOs operate as well on the broader international stage of
water resource policy, some even having consultative status at the United
Nations. The World Water Council, for example, grew out of the 1992 UN
Conference on Environment and Development. It is a multi-stakeholder
organization aimed at managing the world's freshwater resources, stitching
together countless local governments and subsidiary NGOs. The Nature
Conservancy works on an international scale but implements projects on a
local scale. Examples range from restoring wetlands and marshes to help
designing hydropower facilities to minimize impact.

Not surprisingly given the economic centrality of water, business
groups also participate in the process of policymaking. The World Business

Council for Sustainable Development, for example, explores policy concepts associated with water and projects their potential effects on numerous private sector stakeholders around the globe. On a more local scale, entities such as Protect the Flows provide a business perspective on efficiency and conservation within the Colorado River Basin — one of the most heavily engineered and overtaxed rivers on Earth. An important intersection of business interests and water resources is the pendulum swing of privatization of water, which will be addressed in the next section.

Privatization of water

Private water companies have been the subject of intense societal debate in recent years, but they have actually been around for quite a long time. Europe has largely pioneered these ventures, and perhaps the earliest examples come from the British Isles. English and Welsh providers first emerged in the 1600s, mostly representing small service areas and, therefore, there was substantial variability in quality of service due to the fragmented market. By the early 19th century, there were no fewer than six separate entities operating within London alone. British water companies also extended their reach beyond northwestern Europe. In the 1850s, one such company installed the first piped water system and water treatment plant in Germany's capital. Around the same time, France joined the fray, and it has maintained a dominant place in the global picture for well over a century. The Compagnie Générale des Eaux (1853) and Lyonnaise des Eaux (1880), which later became Veolia Environnement and Suez Environnement, respectively, represent the world's two largest private water companies today. On the other side of the Atlantic, the majority of piped water systems in the U.S. were also privately owned during this period.

In the latter half of the 19th century and early 20th century, however, the trend of increasing market share for private water companies reversed in Europe and the U.S., driven in large part by the inability of the companies to expand access to growing populations with greater needs and expectations. Municipal water utilities grabbed steadily larger portions of water service provision, diminishing private operators to 15–30%. In search of new business opportunities, the major European water enterprises looked to Latin America, Asia, and Africa. Privatization of water took hold in major cities in Argentina, Uruguay, China, Lebanon, Egypt, Morocco, Senegal, and other countries. Closer to home, French water companies identified a new pathway to strengthen their foothold within Europe and overcome their relative inability to make major investments: developing contracts in which the municipality owned the infrastructure and held responsibility for investing in its expansion while the private entity oversaw the operation and maintenance. Under this framework, private operation of water systems in France expanded once again, continuing throughout the latter half of the 20th century. A hallmark event in the expansion of water privatization occurred in 1989. The Thatcher administration, promoting conservative philosophy, fully privatized all public water and sewer services in England and Wales (Scotland resisted this effort, maintaining public control over water systems). This move, propelled by the global shift toward a free-market vision, influenced control over water across the planet, including entry into the once Communist eastern European markets such as Romania, Hungary, Bulgaria, and the Czech Republic. Now, in the 21st century, it appears the pendulum is once again shifting direction. Numerous cities are terminating agreements with private operators and returning urban water provision to public control. This municipalization of water systems is global in scope, with key examples in the U.S., France, Latin America, and Africa. Understanding these shifts back and forth through history requires some awareness of the motives that underlie water privatization.

Various stakeholders carry different motivations when it comes to privatization of water systems. In many regions of the world, public water utilities suffer from mismanagement issues such as rampant leaks and inefficient collection of revenues through poor billing practices. Moreover, intermittent service is often a pervasive challenge. Proponents of privatization argue that corporations have sufficient business interest to ensure good management practices, to minimize "non-revenue" water (i.e. leaks), and to expand reliable service to a broader population. Driven by such goals, organizations such as the International Monetary Fund and the World Bank — as well as many regional development institutions — have implemented policies that favor water privatization. The developed world, where access is nearly universal and the quality of service is satisfactory, the motives for water privatization are different. Generally, financial motivations drive such decisions. These can take the form of governments seeking private resources through front-loaded contracts to address short-term budget challenges, with the corporations viewing such arrangements as long-term investments in which they will reap rewards from consumer fees in future years. In a certain fashion, this is a mechanism for a municipality to borrow money against future revenue without having to get public approval for issuing bonds. It is not surprising, then, that one of the complaints that arises following privatization is prices tend to increase — in some cases substantially — as companies seek to update infrastructure and to maximize their profits.

As alluded to previously, privatization can take several forms. Many of the original private water companies undertook full privatization, where the company owns and operates the assets in their entirety. There are still some regions, such as England, where full privatization is the norm following the Thatcher initiative. More common these days, however, are so-called public-private partnerships (PPPs). Under such arrangements, the public continues to own the assets, and specific functions are assigned to the private company for a defined period of time. A PPP can take the form of a management contract where the operator simply runs the water system as a service in return for a fee, which is typically related to performance metrics. Concession contracts lean more heavily on the private partner, involving operation as well as private investment in upkeep and improvements — these are perhaps the most common type of PPP today. Another flavor of PPP, structurally somewhere between a management contract and a concession contract, is a lease contract, or "affermage." In these arrangements, the public assets are leased to the private partner for a defined period, and the private partner collects some or all of the revenues from the system.

Depending on how one defines what constitutes a private water company and whether only water supply or if wastewater treatment is also included, estimates of the privatized market share in today's world vary considerably. Roughly speaking, nearly one billion people are served by the private sector, with the largest fractions in China, the U.S., and Western Europe. Privatization has seen a spectrum of consensus successes to utter failures. Deciphering the impact of privatization in the water sector is indeed rather complex. Ideally, one would compare the changes in various metrics (access, costs, water quality, etc.) when a water utility passed into private management with changes associated with one remaining in public hands over the same period of time and under similar conditions. Such straightforward comparisons do not exist, so proponents on both sides of the issue tend to focus on imperfect studies that draw conclusions aligning with their perspectives. For example, water tariffs tend to rise in the long run regardless of who is managing the utility (see the next section for more on water pricing). When the utility is privately managed this can appear to be corporate greed, and when publicly managed it can appear to be bloated bureaucracy or corruption. While those factors can each play a role, the actual costs to provide the service are driven up by systemic factors related to infrastructure, pollution, scarcity, regulation, and others independent of the management. An important component of the discussion of water privatization, however, is the human right to water. In 2010, with Resolution 64/292, the United Nations General Assembly explicitly recognized the human right to water and sanitation. This resolution acknowledged that clean drinking water and sanitation underlie all human rights and placed the onus on governments and international organizations to deliver financial and other resources to provide safe, clean, accessible, and affordable drinking water and sanitation for all people. Whomever runs our water systems, those systems must stand up to this expectation and deliver one of the most fundamental of human rights universally.

Price of water

If it was not clear before reading this book, by now it should be obvious that water is probably the most vital material on the planet. How, then, can one reconcile that with the fact that water is so cheap that, in some places, it is literally free? This conundrum, known in economic circles as the paradox of value (or the diamond–water paradox, as in: why do diamonds cost so much more than water when the latter is clearly more useful?), has been pondered for millennia. Although origination of the paradox of value is often attributed to 18th-century Scotsman Adam Smith, similar queries were posited as far back as Plato's dialogue *Euthydemus*. One simplistic explanation to resolve this paradox is that diamonds, for example, are far scarcer than water, and scarcity increases value and cost. A more nuanced explanation is based in the concept of marginalism. It is not the general usefulness of water or diamonds that governs price, but rather the usefulness of each unit of those things. Yes, the general utility of water cannot be overstated. However, the overall supply of water is substantial, so it carries low marginal utility. That is, each additional available unit of water can be used to address increasingly less urgent needs and, therefore, carries less value. Diamonds, in contrast, have limited supply. The utility of an additional diamond outweighs that of, say, one additional glass of water. One wanting diamonds is willing to pay a higher price for one diamond than for one glass of water. Sure, a thirsty person stumbling through the Sahara Desert may have more marginal use for a glass of water than a diamond, but only to the point when the immediate need is quenched.

So we have covered why water is so cheap, but just how much does it cost? There is no single answer to this question. For one thing, it depends on the source from which you obtain your water. We will spend much of this section on municipal supplies, but as described earlier in the book, this represents a relatively small fraction of overall water withdrawals. In many cases, especially for agriculture and certain industrial settings, users directly abstract water from either an aquifer or surface source. Since these abstractions do not come from either an independent private or public supplier, monitoring and managing them is a monumental challenge. In some localities a government entity will charge an agricultural user by the area of farmland irrigated, perhaps scaled by the type of crop since some plants require far more irrigation than others. Generally speaking, however, direct abstraction of water incurs no charge to the end user at all — particularly in countries outside the OECD. Even in locations where farmers do pay for water abstraction, the price per unit volume is typically 10–100 times lower than what urban residents pay. Given the substantial share of water consumption associated with growing food worldwide, implementation and enforcement of water pricing for direct abstractions will need to expand and improve. In many ways, these water consumers do not pay for the "externalities" of their water abstraction, i.e. the societal and environmental costs of using this natural resource. Let's turn our attention now to those sectors where pricing is already broadly enforced.

Whenever considering a price for water, it is imperative to keep in mind that water is a human right necessary for survival. It is unethical to price people out of water. One approach to avoid this is to subsidize a minimum amount of water per person, ensuring that the poor have guaranteed access to sufficient water for basic daily needs — approximately 15 gallons per person. Recognizing that wealthier individuals have no need for such a subsidy, an alternative approach is to tie the pricing to income. When considering minimal water needs, it is informative to examine how per capita water consumption varies around the world. Figure 5.3 provides these data for selected countries spanning all of the populated continents. The contrast is stark. Many countries in Africa and the Caribbean have populations subsisting on daily water supplies lower than what is considered necessary for basic needs. The gap between these countries and places such as Australia or the United States is striking and emphasizes both the inefficiencies in much of the developed world and the impact that development and infrastructure can have on human prosperity and health.

Average per capita daily water use (gal)

0	20	40	60	80	100	120	140	160

Mozambique
Uganda
Rwanda
Haiti
Ethiopia
Cambodia
Angola
Nigeria
Burkina Faso
Nigeria
Ghana
Kenya
Bangladesh
China
India
United Kingdom
Phillippines
Peru
Brazil
Germany
Denmark
Austria
France
Norway
Spain
Mexico
Japan
Italy
Australia
United States

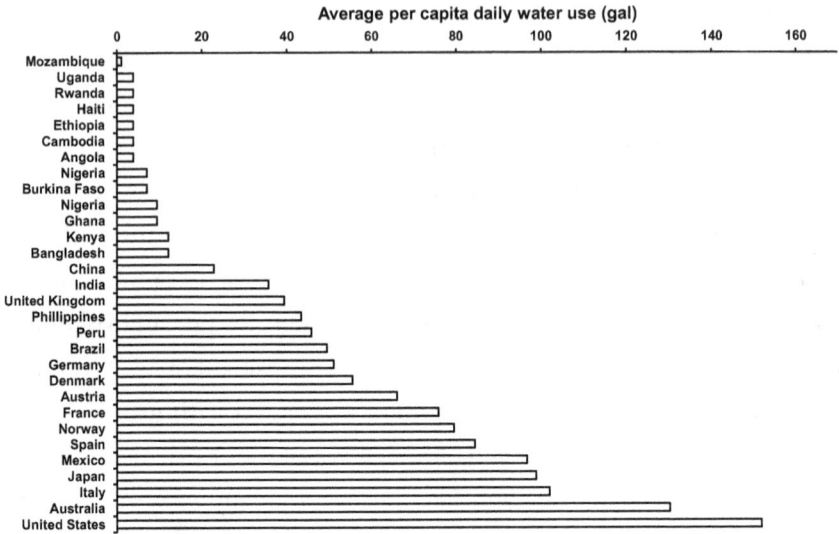

Figure 5.3.   Average per capita daily water use in selected countries (2014 data from data360.org).

Returning to the concept of cheap water, it is interesting to note that, while residential and business consumers do pay water bills, those charges are often based on delivery of the water — the cost to get the water to one's home or business — not on the water itself as a product. In a sense, then, the water is actually free. Furthermore, there are many locations where consumers do not have water meters at all; rather, they pay a flat rate regardless of the volume of water they consume. Not surprisingly, having no meter correlates with wasteful water practices. It also makes price controls powerless as a means of regulating consumer behavior. Penetration of water meters has been growing for this reason. Price is a powerful instrument for water policy. Charges for use by volume are a central tool. For price to work as a tool to drive conservation, the consumer must be aware of it. In the U.S., gasoline prices are posted on stations on thousands of roads. A typical residential consumer receives a water bill either monthly or quarterly. There is often no price or meter to see, so there is no direct signal that conservation saves money. Water-scarce communities are starting to use social messaging, such as through school systems to broadcast messages about water conservation. There can also be charges or fines for polluting water (typically associated with industrial users), with the penalties scaled to reflect how difficult the particular pollutants are to remove from the water. Residential consumers also pay to clean up wastewater in the form of sewer bills, which are often folded into the water supply bill. Indeed, sewer prices often exceed water prices, in keeping with the fact that treating wastewater requires more energy and other resources than treating drinking water. As more municipalities move toward water recycling, that is, direct potable reuse of treated wastewater, these prices will eventually converge. There are limits to the influence of price as a means of encouraging efficiency in water practices. Often, local laws preclude a water utility from charging more for water services than they cost. While such regulations help prevent price gouging, either as a means to fill municipal budget gaps or purely for corporate profits, they also hamper attempts to respond to scenarios such as extended droughts in which consumption by individual users can be difficult to curtail. Another limitation to pricing as a behavior control is that some industrial users may opt to build their own treatment systems to circumvent costs for municipal wastewater treatment of their effluent. A third challenge in tying price to cost in the water industry is that the majority of a utility's costs are fixed, such as debt payments, rather than scaled to the amount of water they deliver to consumers. Since the latter is the primary revenue source, there is a structural mismatch between costs and revenues, which can be exacerbated when consumer consumption varies as a result of (otherwise desirable) conservation measures. In other words, conservation reduces revenue to the water utility while costs remain the same.

Historically, water utilities have exercised so-called decreasing block pricing. This means that the price of water depends on the volume consumed, with lower price tiers for larger volumes — a bulk discount, if you will. While this principle is familiar with many goods, as seen in the big box retail industry, when it comes to water, decreasing block pricing can promote waste and abuse. In recognition of such concerns, water utilities have been gradually shifting toward increasing block pricing to penalize users who consume more than their share of the resource and to cut down on waste. Another trend in water pricing is steadily rising prices in almost all markets. Increasing prices are driven by dwindling supply, the need to pull water from more polluted sources, and — in many areas — the need to maintain and replace aging infrastructure such as water mains that are well beyond their expected life. The backlog of infrastructure needs in the United States alone has been estimated at more than $1 trillion.

Water prices vary dramatically, not only among different countries, but also within a given country. Figure 5.4 provides a sense for the latter by presenting combined water, sewer, and stormwater prices across a collection of cities in the United States. Average monthly water bills vary from below $50 to well over $300, but in almost all cases, the absolute prices in the U.S. have risen nearly 50% since 2010. One thing interesting about this price discrepancy is that it does not closely correlate with rainfall and local water availability. Fresno is in a water-scarce region in California while Seattle experiences more rain than just about every

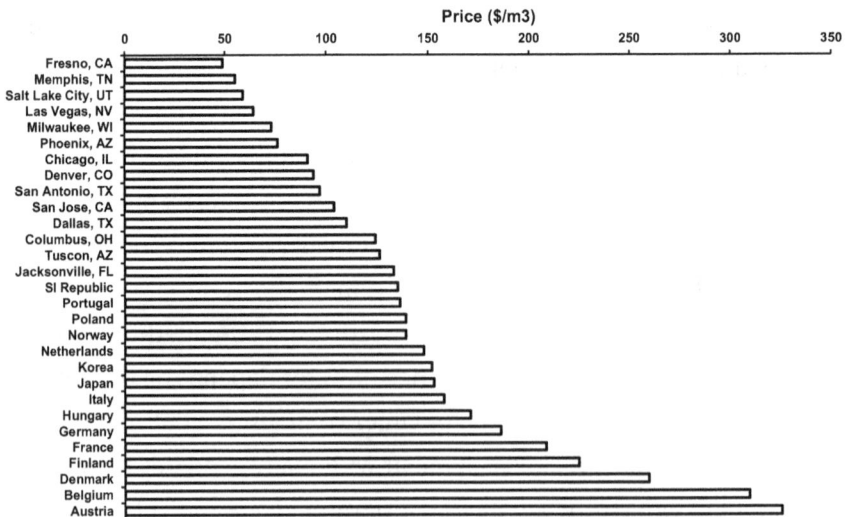

Figure 5.4.   Combined monthly water, sewer, and stormwater residential prices in several U.S. cities (2015 data from *Circle of Blue* survey).

Price ($/m³)

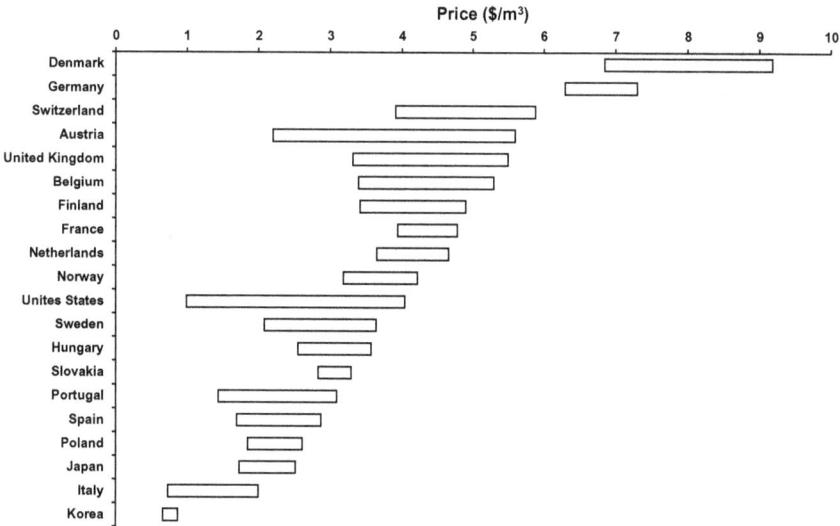

Figure 5.5.   Range of prices for water (in US$/m³) in selected countries around the world (2009 data from OECD).

other city in the U.S. Even with this steep (and ongoing) increase, water prices in the U.S. are comparatively low on a global scale. As can be seen in Figure 5.5, Europe, in particular, has far higher prices for water, with Denmark and Germany representing the most expensive countries. Though not perfect, there is a correlation between these water prices and per capita water consumption in these countries, driving home the point that price can be used to manage behavior. Places where water is cheap, like in the U.S. or Japan or Spain, each person tends to use substantially more water.

Pricing of water is different than energy, the other large commodity used throughout the world. Water prices are local and variable. Natural gas prices tend to be regional or even continental, and petroleum prices tend to be global. One driver is the fungibility of the commodity. We typically only transport water within a region, so each area has its own pricing mechanisms. Natural gas can be readily piped across a continent, but is somewhat limited in transoceanic trade. The transoceanic trade of petroleum in supertankers creates a market where petroleum prices vary only a few percent virtually anywhere where ports and pipelines are accessible. Water pricing is local and doesn't generate an international market. This is one reason why virtual water, discussed in Chapter 4, is an important concept.

One of the most progressive nations when it comes to pricing of water is Israel. Situated in a water-scarce region of the world, the Israelis have invested heavily in water security, including installation of state-of-the-art desalination plants to reduce dependence on the Sea of Galilee, Jordan River, and unsustainable groundwater sources. As a reflection of the substantially different costs connected to producing fit-for-use water from sources as disparate as treated wastewater, desalination, surface water, and aquifers, they have difference prices tied to the source. Israel considers all water to be a nationally-owned asset. Israel charges a farmer the full value of the water, even if they abstract it themselves from a well on their own property. In this way, the market can be employed to assist in optimizing use of a limited resource. For the Israelis and their Middle Eastern and North African neighbors, water scarcity has a firm connection with past, current, and future conflicts. In the next section, we will dive further into water conflict.

Water conflicts

While outright wars are rarely fought explicitly over water, it has been an aggravating factor in literally thousands of conflicts through human history. Following the violence that overwhelmed Rwanda in 1994, many have attempted to understand the origins of the human tragedy known as the Rwandan genocide. The typical explanation revolves around how two rival ethnic groups, the Hutu and Tutsi, were embroiled in a bitter dispute that eventually escalated into the events of 1994. Prior to that time, however, the government in Rwanda instituted environmental programs, and these produced unintended consequences that fostered tensions. A reforestation campaign begun in the early 1990s, for example, placed large swaths of eucalyptus across the nation, which is a particularly greedy plant with respect to water and nutrients. Coupled with other threats to Rwanda's water supply, farm production plummeted, leading to food shortages and, ultimately, violence. The conflict in Sudan's Darfur region claimed hundreds of thousands of lives and displaced two million people. This violence, too, was triggered by disputes over scarce water, in this case between black African farmers and Arab pastoralist communities. More recently, the brutal civil war in Syria was sparked when residents of Daraa, a city on the southwestern tip of the country, became outraged at the corruption of local officials' allocation of limited reservoir water and vandalized property with anti-establishment graffiti. Their subsequent arrest and torture by the government in 2011 ignited extant revolutionary sentiments. Given that about 10% of the world's population lacks reliable access to drinking water, there will undoubtedly be more examples such as these in the years to come.

Examples such as the Syrian uprising represent one type of water conflict, namely, over control of and access to water resources — sometimes motivated by political intentions (such as by offering preferential access to favored allies). Another classification of water conflicts is in the use of water itself as a military weapon or, relatedly, when water systems are targeted by military or terrorist actors. In other cases, water conflicts involve socioeconomic development disputes over water resources. Looking back through history, one can uncover countless examples of each of these categories of water conflicts. Among the earliest references to the use of water as a military tool is found in the biblical Book of Exodus. As Moses led the Jews from Egypt, pursued by the Pharaoh's army, the Red Sea purportedly parted to allow their passage before crashing back in on the Egyptians. Twelve hundred years later, Gaius Julius Caesar put an effective end to the Gallic Wars with the Siege of Uxellodunum. Caesar concluded that the city could not be taken by force, so he decided to target the water supply, which the Gauls collected by descending a steep slope to a nearby river. Roman archers stationed near the river to block access to this source. A secondary water source, however, flowed directly under the walls of the fort. Caesar's forces eventually identified the location of the spring feeding this source. He ordered engineers to build an earthen ramp that could support a siege tower, from which the Romans bombarded the spring, while other engineers constructed a tunnel system to drain the water from the spring. Roman victory was decisive. The Siege of Uxellodunum was the final major military confrontation of the Gallic Wars, marking the pacification of Gaul under Roman rule. Jumping ahead another millennium, denial of access to water resources was also a powerful tactic employed by An-Nasir Salah ad-Din Yusuf ibn Ayyub, known more popularly in the West as Saladin, the first sultan of Egypt and Syria and founder of the Ayyubid dynasty. On July 4, 1187, in the famed Battle of the Horns of Hattin, Saladin defeated the Crusader Kingdom of Jerusalem by luring the army away from reliable spring-fed water sources and filling wells with sand along the way to prevent local villagers from supplying the Christian soldiers with water. The Muslim forces, in contrast, had access to water via camel caravans transporting water from nearby Lake Tiberias. Surrender was inevitable.

Water has also been applied as a defensive military tactic in conflicts. A classic example is the Dutch Water Line, first implemented in the 16th century during a lengthy war between Holland and Spain — and repeated numerous times thereafter. Since much of Holland sits below sea level, intentional flooding of the land was used to break Spanish sieges of the towns Alkmaar and Leiden. A similar tactic was employed centuries later on the other side of the globe by Chiang Kai-shek, then the chairman of the National Military Council of the Nationalist Government of the Republic of China. Kai-shek ordered the dynamiting of dikes used to control Yellow River flooding, spilling water across the plain and miring the invading Japanese army in mud. Explosive destruction of water systems has also been a hallmark of various instances of domestic terrorism and acts of protest. In April 1887, residents in Paulding County, Ohio blew up a canal reservoir, believing it to be a public health hazard. In the early 20th century, the Los Angeles Valley aqueduct and pipeline were repeatedly bombed by those angry over diversion of water from the agricultural Owens Valley to the growing city. Governments have applied related strategies as an offensive military tool. Aerial bombardments of water systems — and dams in particular — have been a familiar occurrence in wars through the latter half of the 20th century, starting with World War II and continuing throughout the Korean and Vietnam Wars.

Today, too, there are numerous examples of brewing interstate conflicts involving water. Many of the most visible such conflicts revolve directly around access to water resources where they cross international boundaries. The Tigris and Euphrates Rivers emerge from sources in the Taurus Mountains of eastern Turkey, from where they flow through Syria and Iraq into the Persian Gulf. Climate changes have reduced the flow of this river system, but there have also been increasingly large withdrawals upstream, further limiting the amount of water reaching lower elevations. Given additional planned development of hydroelectric resources in Turkey, this is a recipe for significant conflict in the near future. Nearby, the Nile River faces an analogous fate. The Blue Nile starts in Lake Tana in the Ethiopian Highlands and flows to Khartoum, where it merges with the White Nile, itself coming from Lake Victoria and flowing north through Sudan. Together, the Blue Nile and White Nile become the Nile River. Recent and planned development in Sudan and Ethiopia, including major dams, will have dramatic impact on the water resources available downstream in Egypt. In Southeast Asia, there is yet another comparable example with the Mekong River, where dams constructed in upstream Laos will negatively affect Cambodia and Vietnam.

The Israeli–Palestinian conflict is among the most infamous globally, yet many outside that region are unaware that water is central to the struggle between these peoples. In the 1960s, the Israelis constructed a massive water engineering system known as the National Water Carrier. This system transfers water, at one point nearly two million cubic meters each day, from the Sea of Galilee in the north to the populated center and arid south of the country. (Recently, flow rates have decreased as Israel expands production of desalinated seawater and recycled wastewater, offsetting demands on traditional surface water sources.) This water would naturally have flowed through the Jordan River, which is a critical source for Syria and Jordan. Diversion of water from the Jordan River has therefore been a longstanding source of tension in the region. In 1964, Syria attempted the construction of a system to reduce the capacity of the National Water Carrier. Israel's military attacked to stop this effort the following year, which was a major contributing factor initiating the 1967 Six-Day War. In that conflict, Israel captured the Golan Heights from Syria, which it occupies to this day. Strategically, this region is significant because it houses sources of the Sea of Galilee. Newer national efforts by Israel to reduce water consumption and to increase supply with desalination plants and wastewater reclamation and reuse have mitigated these historic tensions to some degree. However, there is still controversy surrounding the notable disparity in per capita water access between Israel and Palestine. Another long-running interstate conflict with water as a central factor is that between India and Pakistan. The Indus River provides water resources that are pivotal for Pakistan's economy, both for the agricultural breadbasket of Punjab province and for many heavy industries. It also represents the primary source of drinking water for the country. Some of the headwaters of the Indus River are located in Kashmir, the hotly disputed region at the border between the two countries. There are numerous cultural and historic underpinnings of the strains between India and Pakistan, but water is surely among the most powerful.

In Central Asia, another water conflict is brewing. The Rogun Dam under construction on the Vakhsh River in southern Tajikistan is intended to produce several gigawatts of electricity for the Tajiks' aluminum production, but downstream Uzbekistan has expressed dire concerns for the fate of its lucrative cotton fields if flow rates in the Vakhsh are reduced and irrigation becomes impossible. Irrigation in this region has already taken a major hit with the reduction of the Aral Sea, situated between Kazakhstan and Uzbekistan. The Aral Sea was once among the largest lakes in the world, but it has progressively shrunk starting in the middle of the 19th century. This shrinking accelerated significantly in the middle of the 20th century as a result of major diversions of its feeder rivers (the Amu Darya in the south and the Syr Darya in the east) for irrigation. It is now a mere shadow of its former self, maybe 10% of its pre-diversion volume, split into several shallower lakes, representing a true environmental disaster. Fishing, previously a reliable industry, is now virtually gone. Satellite images of the lake over a 40-year time span are frequently used to demonstrate the environmental impact of unsustainable human endeavors. Kazakhstan, Uzbekistan, Turkmenistan, Tajikistan, and Kyrgyzstan are all embroiled in disputes over the remaining water resources.

Fishing in the world's oceans has frequently arisen as a source of international conflict. Brazil and France famously become entangled in the Lobster War in the 1960s. The Brazilian government banned French fishing vessels from catching spiny lobsters 100 miles off its coast. Brazil made the (now comic) argument that lobsters crawl along the continental shelf, while the French maintained that lobsters swim, meaning they can legally be caught by a vessel from any country. Confrontations over fishing rights have also erupted several times between the United Kingdom and Iceland over cod in the North Atlantic, between Canada and Spain over halibut near Greenland, between the British Royal Navy and French fisherman over rights near the Channel Islands, and many others. An emerging conflict is developing in the South China Sea. China is building habitable islands from uninhabited reefs. They are using the new islands for justification of sovereign control of the waters between mainland China and the islands. The countries around the Sea dispute this sovereignty. In such cases, since commerce is central, the World Trade Organization can often serve as an intermediary to assist in resolving disputes. This is also the case for international conflicts surrounding government subsidies for water-intensive goods — especially crops — that are then exported, representing virtual water trade. Generally speaking, it is conceivable that interstate conflicts over water will be less common in the future thanks to the sorts of agreements described at the beginning of this chapter. This does not mean, however, that conflicts in general will abate. On the contrary, those between ethnic groups (as in the cases of Rwanda, Darfur, and Syria) or competing economic interests (as in the Owens Valley and Los Angeles aqueduct) will almost assuredly multiply and intensify as demand continues to rise amidst dwindling supplies of fit-for-use water.

06 Water is Innovation

Chapter

Water infrastructure in the modern world

Societies have continuously advanced how we secure, purify, convey, and treat water. While invisible to virtually all residents — except when there is a crisis — every developed community has installed capacity for producing and delivering potable drinking water and water for other needs, as well as treating and discharging wastewater and storm water. In Chapter 4, we described much of the history of water through human civilization. We start this chapter by describing how today's developed world manages water.

In metropolitan areas, most residents receive water from a publicly owned and operated municipal water utility. Some utilities are operated by private companies, but they operate in essentially the same fashion as the public entities. Freshwater is typically extracted from a surface water source, including lakes and rivers. Many metropolitan areas dam a river and create a reservoir, or essentially a human-made lake. Whether from a lake, river, or reservoir, this water comes from seasonal rains or melting snow from the winter, in certain areas from glaciers, and even from underground springs that reach the surface. In most cases, the water is collected from higher elevation, and only gravity is required to convey the water. When water is collected from a lake at lower elevation, such as the Great Lakes, it must be pumped to treatment plants. Depending on the location and volume, water is conveyed in either open channels or enclosed

123

within large pipes. Open channels are cheaper than sealed systems but increase evaporative loss and increase the risk for contamination. We transport and treat more water than any other substance on Earth — by far. As a comparison, the U.S. consumes about 800 million gallons of oil per day, whereas the Jardine water plant on Lake Michigan alone provides Chicago and the surrounding community with about 900 million gallons of water per day. While volumes are gargantuan, prices for water are extremely low, so revenue to treat water is limited. For example, while oil prices are about $2 per gallon, water is priced at about $2 per 500 gallons. Therefore, utilities are extremely cost-conscious.

In the purification plant, a series of inexpensive but effective steps are taken (Figure 6.1). First the water is passed through screens to block fish, plants, and miscellaneous debris. Some disinfectant is used at the start of the treatment train to reduce risk of algae or other biological growth in the plant. Next, water is treated to remove large suspended particles including oils, metals, and biological species. Frequently, coagulants, which are chemicals that ball up in the presence of water, are used to trap suspended solids and form "flocculants" (Figure 6.2). The water is aerated because the air bubbles stick to the flocculant and cause them to float to the surface, where they can easily be skimmed off. After flocculation, the water is treated by sand filtration to trap flocculants that managed to escape skimming and other smaller materials remaining in the water. The sand must be cleaned periodically, so there are always multiple filtration units, with some offline going through a cleaning cycle while others are in operation. These sand filtration units are kept in the dark to limit algae growth. Just before that water leaves the purification plant, it is treated with chlorine to disinfect. The utility cannot control the time between treatment and customer usage, so the chlorination is designed to last several days. If you don't use your faucet for several days, it is recommended that you flush your water lines, because the disinfectant will lose activity and the water has some risk of recontamination. Municipal water in many locations is also treated with fluoride to strengthen teeth and prevent tooth decay. (Fluoride does not provide any disinfection

Source water treatment

Primary treatment

Screening	Primary disinfection	Coagulation	Sedimentation	Aeration/ Flotation

Secondary treatment

Sand filtration

Tertiary treatment (varies by end use and location)

Ion exchange	Reverse osmosis	Adsorption
Ultra/nanofiltration	Polyphosphates (for lead pipes)	Disinfection
pH adjustment	Fluoridation	

Wastewater treatment

Primary treatment

Screening	Primary settling

Secondary treatment

Bioreactors	Secondary settling	Nutrient recovery

Tertiary treatment

Disinfection (UV, chlorination, ozone)

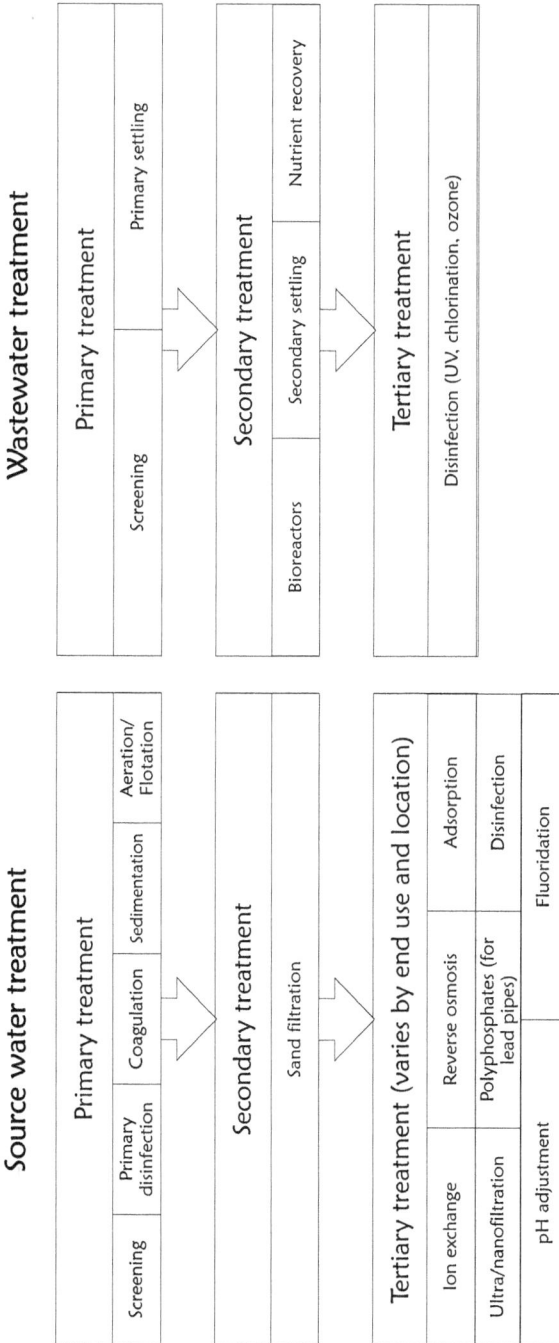

Figure 6.1.   Steps in water treatment plants.

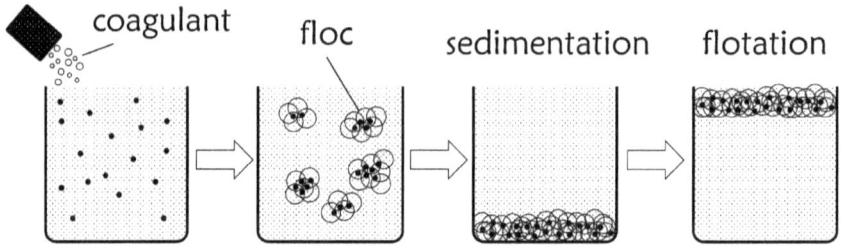

Figure 6.2.   Coagulation-flocculation process.

benefit.) In the U.S., the Safe Drinking Water Act was enacted in 1974 and requires utilities to monitor and report on water quality and contamination, including bacteria, nitrates (from fertilizer runoff), toxic metals, and organics. There is growing interest to prevent drugs, steroids, and other biologically active chemicals from entering the water supply, but inexpensive options are limited for either detection or removal.

After rendering the water potable in the plant, it is conveyed through underground pipes to customers. Pipe diameters as large as five feet leave the treatment plant and as small as a few inches enter individual homes or apartments. Water pressure is required and is either provided by underground pumps or water towers. The central pipes are typically made from cement or iron while the pipes in your home are often copper or PVC (polyvinyl chloride). As water systems age, leaks develop. In many communities, as much as 50% of all treated water is lost to leaks. This "non-revenue" water increases per-gallon costs to consumers and represent an obvious target for improving sustainability of our water systems. Lead pipes have also been used since the Roman era. In fact, the symbol for the element lead on the periodic table is Pb, derived from the Latin *plumbum* — the root of the English word "plumbing." While lead pipes were banned from new construction in the U.S. in 1986 due to concerns related to impacts on human health, water service lines are used for generations, so there is still extensive use of lead pipes throughout the world (i.e. water service pipes have sections that were built before the lead ban). When water is delivered in these old lead service lines, chemicals are added to the water to coat the lead pipes' inner wall and minimize the release of lead into the water. In the case of Flint, Michigan, the state switched from a more distant supply from Detroit to the local Flint River. The water chemistry from the Flint River was different enough from the previous supply that it caused the protective coating to become dislodged, and large amounts of lead was introduced into the water supply to catastrophic effect.

At the end of the pipe, customers frequently use additional in-house water treatments. Under the sink or water pitcher units include filtration, ion exchange, or sorbent units to capture trace amounts of lead, other metals, and biological species. These systems are effective but are far too expensive to treat the entire water supply.

Water from underground sources are frequently considered "hard." Hard water is high in calcium and other minerals with limited solubility. Hard water, while not a health hazard, forms white coatings on surfaces and also makes detergents less effective. Hard water is treated by either ion exchange, which swaps the hard minerals for common salt, or reverse osmosis, which removes the minerals. Softened ion exchange water can taste too salty for many people, leading them to resort to bottled water for consumption. Water is needed in industry as well, and process water for the pharmaceutical industry or the semiconductor industry may require additional purification beyond the municipal water provided. The most common requirement is to reduce salinity far lower than required for drinking water. This polishing is frequently accomplished with a treatment train that includes either distillation or reverse osmosis, a pressure-driven process, followed electrodeionization, an electrically-driven process.

We are all quite familiar with water appliances in the home: the sink, tub, shower, toilet, washing machine, and dishwasher. Over the past generation, we have seen a continuous improvement in the water efficiency of these appliances. Toilets use less water per flush. Showers use a finer droplet size, reducing the amount of water. Washing machines that load from the front consume less water per load. Dishwashers spray more efficiently and use much less water. We also have reduced leakage from faucets and valves. (While these advancements increased water efficiency, a pernicious new problem has developed in some buildings. Legionnaire's disease is caused by legionella, a bacterium that grows in room temperature, stagnant water that is not exposed to the air. Water-efficient plumbing in which the chlorine has lost activity is an ideal environment for growing legionella. Legionella infects us through our lungs by inhaling aerosol containing the organism while we stand in front of the running faucet.) After we use water in our homes, schools, or businesses, we send the now-dirty water down the drain — and most of us think nothing more of it. What happens to all that polluted water?

Wastewater treatment is similar to providing potable water, but in some ways, it's in reverse. We convey and treat the water, again with little room for expenses. While consumers may be willing pay $2 for 500 gallons of water, they are willing to pay approximately $0 for managing wastewater. We mentioned earlier that the drinking water plant in Chicago treats and conveys a larger volume of water than the entire oil sector — well, wastewater treatment plants typically manage twice the volume of the local water utility. The larger volume arises from the inclusion of storm water. The wastewater process starts with release from the home. Six-inch cement or PVC pipes convey the wastewater from the home to the sewage collection line under the street. A sump pump ejects storm water that collects under homes, and we use an ejector pump if we use water in our homes at an elevation below the sewage lines that transport the water to the collection pipe in the street. Gravity, thankfully free, is the only energy source used to convey wastewater to the treatment plant. Wastewater is transported in large cement pipes, six-feet or more, toward the wastewater treatment plant. As the water collects, it moves into lines that are buried increasingly deeply underground, providing a downhill ramp for the water. Storm water enters the wastewater line (in combined sewers) or a storm water line from the street surface via grates along the gutter.

When the collected water gets to the wastewater treatment plant, it goes through a series of steps to recover some value, reduce waste handling, and reduce risk to health and the environment. Wastewater can be high in pathogens, so protecting health is always the highest priority in wastewater management. As with drinking water, wastewater is treated through a treatment train that provides different services. The first, or primary, step is sedimentation or sand filtration. We let the crap settle. The settled or filtered water is then run through a biochemical oxidation in the secondary treatment step. Aerobic bacteria (microorganisms that live in the presence of oxygen) are used to break down organic waste, generating carbon dioxide in the process. Aeration — to mix the water around and to ensure the bugs get their oxygen — requires pumping large volumes of air into the wastewater and is by far the highest energy consumer in the process. At this point, the water may be "polished" by running through a fine filtration to adjust acidity, remove trace chemicals, or other needed steps. The water can still contain pathogens and must therefore be handled with care. Except for a few places in the world, virtually all wastewater is ultimately discharged to a natural water body such as a river or the ocean. Therefore, to reduce environmental risk, it has become common in the developed world to disinfect the water before discharge. Disinfection can be accomplished with chlorine, ozone, or ultraviolet light. In all cases, many pathogens are destroyed by this tertiary treatment, and the risk to the environment and human health is significantly reduced. In urban situations, what we consider a "natural" water body may in fact be discharged, treated wastewater. Approximately 85% of the flow in the Chicago River is actually treated wastewater. Starting in 2017, the Chicago region implemented tertiary disinfection treatment on wastewater discharged into the Chicago River system in an ongoing effort to clean up the area's waterways. The goal is to transform the Chicago River from essentially a sewage conveyance channel back to a swimmable water body. (For more than a century, the Chicago River has been diverted to the Sanitary and Shipping Canal to protect drinking water supplies drawn from Lake Michigan.) While the wastewater is treated, the sludge collected at the beginning of the treatment train is injected into anaerobic digesters, producing organic acids. The generated acids slow down the biological activity and would eventually kill the culture. Therefore, the digesters also support a second class of microorganisms called methanogens. Methanogens consume the organic acids and produce a mixture of methane (natural gas) and carbon dioxide. Energy in the biogas is derived from the carbon in the waste and is ultimately derived from the food we grow and consume. The latent energy in the biogas is actually greater than the total energy consumed in the wastewater treatment plant. Therefore, wastewater treatment plants have the potential to transition from net energy consumers to renewable energy producers! In the anaerobic digesters, some of the carbon is not readily digestible and settles to the bottom of the tank along with any mineral present in the sludge. This material, called biosolids, is blended with soil to improve mineral content, add fertilizer nutrients, and enhance water retention in the soil. Societies have used untreated biosolids as a primary soil amendment and fertilizer source for millennia. We are just returning to that practice now. Other classes of water may require more or less treatment. Storm water generally requires less treatment than residential wastewater, but industrial wastewater may require additional treatment to reduce loads of toxic metals or organic compounds. In cases of highly contaminated industrial water, the wastewater might be injected deep underground.

Of course, not everyone lives in areas with sufficient population density or wealth to warrant the existence of a public water supply. In rural areas, residents typically extract water from wells that provide water to either a single home or a small community of homes. Well water is frequently hard and may be treated with in-home units. In addition, in rural areas, well water is frequently high in nitrates from fertilizer infiltration. Nitrates interfere with hemoglobin in fetuses and newborns. Nitrates are difficult to remove from water, so the most cost-effective solution is frequently to provide bottled water to at-risk babies and pregnant women. Wells are dug deep enough to access the underground water table. Shallow wells are cheaper but increase risk from pathogens, nitrates, and other contaminants present near the surface. Over time, a well can extract water faster than natural infiltration recharges underground aquifers. When water tables drop lower than the existing well, then new wells must be dug — at increased costs. There are many cultural references to the well going dry. Wastewater is also handled differently when public systems are not an option. Homes that are not connected to a public sewer generally use a septic tank. This is a buried chamber made of fiberglass, concrete, or PVC. Waste from the home settles in the tank, and anaerobic digestion reduces the amount of organic waste. Waste that is not decomposed by digestion must eventually be extracted from the septic tank, lest it fill up and overflow (ick). Accumulated sludge is pumped out of the tank by a vacuum truck. Anaerobic decomposition is then restarted when the tank is refilled. Away from municipal utilities, wastewater from septic tanks is treated in septic fields. Septic fields are installed downstream from the well or drinking water source and are simply filtration fields that collect waste and provide an environment for microbial treatment. The water is filtered and slowly infiltrates the soil and recharges the aquifer.

Now that you have a handle on how we deal with water today, at least in the developed world, let's take a look at where things are heading.

Invention, innovation, and intellectual property in the context of water

If water is innovation, the first question is what is innovation? And what is the difference between invention and innovation. Bear with us for a brief digression on this subject so we can all be on the same page. To most people, invention and innovation are almost interchangeable concepts that suggest the creation and implementation new ideas. In this era of interconnected smart phones, innovation is typically associated with Apps and new interactive tools. In the postwar second half of the 20th century, innovation implied new devices such as televisions and microwave ovens. Neither era of innovation suggests that water is innovation.

"As seen on TV" inventions are devices that are conceived to solve a specific problem, designed, manufactured, and sold to customers.

Invention and innovation also suggest research publications in scientific journals, scientific books, and patents. Someone outside of technology fields does not necessarily consider processes, methods, strategies, copyrights, or trademarks to represent invention and innovation. In this section, we define our terms for invention and innovation and highlight how water can be viewed through a lens of innovation. First and foremost, we must consider intellectual property. The word "intellectual" suggest thoughts or ideas and the word "property" suggests something of tangible value. Tangible property can be created, owned, bought, sold, shared, traded, borrowed, given away, destroyed, or stolen. Intellectual properties are those thoughts and ideas that have tangible value, and society enables the owners to be compensated for their economic value. For example, our publisher copyrighted the words in this book. The authors (us) have made an effort to create value (the ideas in this book), and we hope the readers (you) compensate us for our intellectual property. Books, arts, and music are all forms of intellectual property or "IP."

But what is invention and innovation? Technologists and investors wear an economic hat to define invention and innovation. Somewhat offhandedly, invention is the process of converting money into ideas or IP. Basically, invention is the process where we invest in research and develop novel ideas. While novel, those ideas may or may not be useful. Innovation is the process of converting ideas and IP into money or other value. In other words, innovation is taking those invented ideas, creating a useful product for society, and selling it or implementing it at scale. So, it is invention and innovation together that benefit society. All innovation requires invention but not all inventions result in innovation.

With this framework in mind, let's bring this back to water. Researchers develop conjectures about water's behavior, investigate water's properties, and develop theories on the fundamental implications of these discoveries (see Chapters 1–3). Water researchers go on to propose, conceive, design, and develop solutions to water's challenges and opportunities. Both the discovery of properties and design of solutions are intellectual property. The first impression that IP means devices "as seen on TV" is a reductionist view and represents only a fraction of IP, invention, and innovation. Intellectual property includes those devices that are conceived and created. Those devices include new sensors to detect the properties, quantity, and quality of water. Those devices also include membranes and filters to treat or purify water. (As practitioners, the authors spend much of our waking day considering devices to purify water.) The history of water devices is discussed in Chapter 4, and the future of water devices is discussed in Chapter 8. Much of the history was based on society's need to provide and convey water for drinking and agriculture and dispose of wastewater and storm water. That history represents a merging of new devices with major civil infrastructure projects. The historical backbone of innovation in water has been new ways to efficiently convey water to or away from its desired targets. These innovations go back millennia. Only in the 19th century did innovation in water start focusing on reducing contamination and improving purity. The future of water devices is more focused on ways to improve the sustainability and resilience of our water systems. The future of water is focused on providing water while minimizing the impact on the environment and the biota, reducing pollution discharge, reducing energy use and greenhouse gas emissions (sustainability) while building and designing systems to withstand catastrophic failure, whether natural or human-induced (resilience). We will dive deeper into this future in Chapter 8.

With water, IP implies much more than devices. Intellectual property implies new models, concepts, and algorithms to predict stress points in water systems. Intellectual property implies controls systems that rapidly synthesize large sets of data, consider both theoretical and empirical models, and design new decision strategies and control systems. In a nutshell, the emerging fields of big data, high performance computing, and machine learning are starting to impact water.

Innovation can also include social tools. When you remind people of the impact of their actions such as wasting precious water, you can influence behavior and conserve water. Such social tools, including pricing mechanisms, are essentially innovation. In the power grid, we are starting to use smart electric meters and charge customers more to consume electricity at times of high demand. While the water grid is not as dynamic, many areas charge customers more per gallon when their use goes up. This creates an economic incentive for conservation. These innovation tools are becoming more prevalent as we experience more frequent extreme droughts and water shortages. In reverse of water demand, innovation is developing in storm water and flooding situations. We are using regulatory, social, and financial tools to incentivize reducing the use of impervious infrastructure (gray) and replace it with porous infrastructure (green). These innovations can depend on local devices such as rain barrels to collect water from roofs or replacing parking lots with porous bricks that let water soak through rather than run off.

As suggested, innovation in water is strongly dependent on laws, regulations, policies, and permits. Arising from potential risks to health and the environment, water is highly regulated. To implement new technologies or controls requires extraordinary amounts of testing evaluation, and validation. In regions that experience wide ranges in temperature or large variability in rainfall, most water technologies require a minimum of one year, if not two years, of validation to ensure that they work across the range of potential environmental conditions. Implementation of water systems is also dependent on local conditions, such as how subsurface porosity affects hydrology. Therefore, most permitting agencies require that water utilities validate technologies at their specific location and cannot use ancillary results from a similar facility elsewhere. The authors have an ongoing joke that just because a water technology works in Vermont, doesn't mean that it will necessarily work in New Hampshire, and therefore it must be validated in both states.

One of the most challenging aspects of innovation in water is the way society supports water infrastructure and prices water. Access to water is considered a human right. As a human right, we balance the price of water between what people can afford and what it costs society to provide the water service. This model leaves two major challenges. First, we put no inherent value on the water. We assume nature provided the water for free, and the costs are associated with the marginal or incremental expense to convey, treat, and deliver it. As we discussed throughout this book, we are rapidly depleting water resources in many regions of the world. Our pricing models do not address this challenge. Thus, we have created a sustainability gap in water management. Another factor is the age and suitability of our water infrastructure. Most water infrastructure is publicly owned. Without a profit/loss driver, we continue to use existing infrastructure well beyond its useful life. Most public entities must issue bonds to secure capital for infrastructure projects. Bonds typically require voter approval. With a mandate of "if it ain't broke, don't fix it," we let water infrastructure age beyond its expected life. We use patchwork repairs to maintain the water systems as long as possible. We only do major upgrades and replacements when the infrastructure has had a major collapse or failure. Frequently the collapse may have been caused by a natural disaster such as a typhoon, tsunami, or earthquake. Others might be anthropogenic, including a fire or explosion at a nearby facility. Recently there has been growing concern that water infrastructure could be disrupted by terrorist or cyber-attacks. Indeed, the potential for cyber-attacks is growing as we move toward integrated sensing and control systems.

When a community loses access to water or is flooded with wastewater or storm water, it is the wrong time to explore innovations that could improve performance and lower costs. After catastrophic failures, we look for off-the-shelf solutions. Just when the system is most poised for innovation, society often doesn't believe that it is the right time for innovation. Without the profit motive, safety and reliability outweigh performance and cost. This disconnect has retarded innovation in water for generations.

As we discuss in Chapter 8, there are many factors changing our relationship with water, and these bode well for innovation. The stress associated with climate change has increased risk for water availability and impact of flooding. Our increasingly robust models reveal the location and scope of these risks. We can apply economic models and determine current costs for implementing innovation now vs. delaying investments. Basically, we are improving our ability to put a value on current investments. Our enhanced modeling correlates well with global recognition of the benefits of improving the environment and reducing anthropogenic impact. Internet access to images and videos of human and climate impacts on water systems, such as the hurricanes and flooding across the Eastern U.S., forest fires across the Western U.S., extreme flooding in major urban areas around the world — and water shortages in others, and water contamination in Flint, Michigan has changed public perception regarding the need to invest in water innovation.

At the dawn of civilization, water innovation enabled society to provide water to an increasingly dense population for drinking and agriculture. In the 19th century, we learned about the risks of contamination and used innovation to purify drinking water and remove wastewater. In the 20th century, we achieved a level where we could securely provide water and convey wastewater in developed countries with little risk for health. We became complacent and ignored innovation in water. In the 21st century, we are recognizing the impact of our activities and are starting to rediscover our roots in water innovation. Much of society sees both the need and benefit of innovation in water. This holds true from California to China to the Middle East and Europe. Starting around 2010, there has been a growing inclusion of water technologies and solutions in technology investment portfolios, suggesting a recognition that there is value in water innovation. In addition, there is strong growth in water research at universities, government laboratories, and non-profit organizations around the world.

07 Water is Culture

As water is integral to life, it is integral to culture and the arts. Water infiltrates music and songs, poetry, stories, and fables. Water is represented in paintings and other visual arts. Water imagery appears throughout history, from prehistoric drawings to avant-garde theater. A full exploration of water's permeation throughout human culture would consume at least one full book, but in this chapter we will touch on several topics to expose the diverse breadth of water in so many arenas that are quintessentially human.

Water as cultural angst

Native American tribes from the Wabanaki Confederacy of the eastern seaboard have a legend about a frightening beast named Aglebemu. He was said to be as tall as a pine tree, with a fat, green body and a frog-like head large enough to swallow people whole. Aglebemu was a greedy monster. He lived in the river (usually told in the context of the St. John River in New Brunswick or Penobscot River in Maine) and wanted all the water for himself. To prevent anyone else from using his beloved water, Aglebemu constructed a mighty dam to block the river's flow. Fish died, along with the helpful water sprites that lived in the river's depths. People downstream begged for him to release some water, lest they die of thirst and their crops wither away, but he refused. Ultimately, Aglebemu was defeated and transformed into a bullfrog by the hero Glooskap.

Across the world, the indigenous San people of the Namibian Kalahari Desert tell the story of how the zebra got its stripes. There once was a conceited baboon, who declared himself the "Lord of the Water." He selfishly protected a small pool, a rare and precious source of water that remained during times of drought, preventing any of the other animals from quenching their thirst. One day, a zebra and his son came to this pool during a particularly dry and hot period, desperate to partake of the pool's water. The baboon angrily threatened them from his nearby camp fire, declaring the water his and his alone. "Water belongs to everyone," declared the young zebra. At that, the baboon attacked him, leading to a long and ferocious battle, ending with a violent kick from the zebra that sent the baboon flying through the air, landing on some rocks so hard that he lost the hair from his bottom — evidenced to this day by baboons' patches on their rear ends. The kick was so powerful, that the exhausted zebra staggered, falling through the baboon's fire and scorching his white coat. The zebras ran away in fear to the plains, carrying black stripes from the burns.

Stories such as these can be found in innumerable cultures that survive in arid climates or that have faced severe droughts. Lack of drinking water or irrigation is a fundamental threat to life and it is expressed in the art of all societies. When water is plentiful and clean, it returns to the back of our minds, but the stories serve as reminders of the threat. Just as too little water can drive cultural angst, so, too, can too much. Flood legends pervade human civilization, from the Great Flood of Gun-Yu in China to Noah's Ark and the Great Flood to the tale of Utnapishtim from the Sumerian Epic of Gilgamesh. These fears can even manifest so powerfully in some individuals that they rise to the level of a phobia. Antlophobia, drawing its name from the Latin word for water pump (*"antlia"*), is the fear of floods, and it can drive its sufferers to live on upper floors of buildings and, in severe cases, even to living only at high elevations.

Water in language

Water flows through our languages. An ongoing routine among researchers and folks in the water industry is to inject water analogies into our discussions and presentations. With water-related words and expressions filling the language to the brim, it probably happens by accident at least as often as to introduce some humor. Try it, it's easy. Stopping once you start? Not so much.

There are common water words that appear in our stories and poems describing natural water, such as the ocean, seas, rivers, lakes, ponds, streams, creeks, puddles, rain, snow, fog, hail, sleet, ice, steam, slush, swamps, and storms. Some water words describe action and movement, including waves, currents, spray, and torrents. Other words describe our devices and implements, including buckets, hoses, boats, and rafts. Many expressions refer to the world we built around water, such as wells, dams, bridges, tunnels, canals, and levees. Others describe our tools for drinking water, including cups and vessels. When we are in trouble we are in hot water, and when we are emotionally distant we are as cold as ice. When we have plenty, we are awash with good fortune, and when we are on a steep learning curve we are drinking from a firehose. Hope springs eternal. Below is a short piece one of us wrote, pushing the limits of density of water references so you can see things can get out of hand:

Troubled waters ahead, and science may just save us

We are definitely in over our heads. Our unsustainable use of the planet's most important resource will leave us dead in the water. This isn't just about us having enough safe water to drink. The ripples of our looming water crises extend out to all aspects of society. Once you look below the surface, you'll see that the flood gates open to impacts you may never have considered.

We all interact with water in so many ways every day, from showering to drinking to doing the laundry, but residential consumption is actually just a drop in the bucket in our overall water use. In this country, electricity production actually represents the high-water mark for withdrawals of water (it is used to cool power plants). Agriculture is also extremely thirsty with large-scale irrigation, but did you know that manufacturing can't happen without water? It takes water to make anything and everything. We've been spending water like, well, water. We take it for granted, because things seem fine. Water has been abundant, cheap, and safe — at least here in the developed world. But still waters run deep. The truth is that we are relying heavily on groundwater, pumping it out far faster than it is refreshed. The idea that we can keep drawing on groundwater like we have doesn't hold water. With skyrocketing demand and sinking supplies, we are in deep. How could we have all been so stupid as to get ourselves into this predicament? That's water under the bridge now. The question is how we stay afloat.

While smart policies and good old-fashioned leak repairs can make a huge impact, we are going to have to venture into the backwaters of science and technology to blow this challenge out of the water. Feel like a fish out of water when it comes to techno-speak? I know, I know, science…dull as dishwater, right? You and science are like oil and water? Let's pour some cold water on that perspective. Try letting these ideas wash over you and just go with the flow. We won't even need to water down the concepts. Your mouth will water at the prospect of a sustainable water future. Let's gather around the water cooler and dive right in.

Water can get pretty gross. Scientists are coming up with new ways to test the waters, literally, with tiny sensors that can detect trace amounts of a vast array of pollutants in real time. Many of those pollutants are troublesome even to the water systems themselves, sticking to the things engineers are using to clean the water, like membranes, and fouling them up. New materials are being developed that can actually clean themselves, so that the gunk is destroyed or simply slips away like water off a duck's back.

Speaking of wastewater, researchers are realizing that the stuff in our sewers isn't actually waste at all (well, not all of it). Rather, wastewater is chock full of valuable resources like phosphorous and organic matter. We've been throwing the baby out with the bath water all this time. Researchers are developing new ways to extract those resources out of waste streams with things like algae and even highly selective sponges. These ideas are just the tip of the iceberg. There are so many ways to innovate in this field that scientists can smell blood in the water, and their ideas are coming in waves.

Still not convinced this is perhaps the biggest challenge we face as a society? Well, I guess you can lead a horse to water, but you can't make it drink. There are uncharted waters over the horizon, but come hell or high water, science can help us make a splash. Let's not muddy the water by getting bogged down in international agreements and water rights. Discovery and innovation are the rising tide that lifts all ships. Join us as we forge a sustainable water future for all humankind. Come on in, the water's fine!

Water in the arts

Water themes are woven deeply into poetry, literature, theater, film, music, and fine art. While representations of water are at times literal, they often serve as a metaphor for self-discovery, change, placidity, violence, and renewal or rebirth. Water is mutable. It is transcendent. Water is both destructive and sustaining. It transports adventurers and strands castaways. It represents at times freedom and at others imprisonment.

The Odyssey, the epic poem representing one of the earliest examples of Western literature, describes Odysseus' perilous decade-long journey across the sea following the Trojan War. Jumping forward nearly two millennia, in the Far East, the Ming Dynasty era story of the Ten Brothers draws deeply on water. It has been reimagined in many forms, but in most incarnations the youngest brother has the magical power to swallow or cry a river or the sea. The outcome of the story frequently comes to the brothers' adversaries drowning. In the early

18th century in England, one of the earliest and most influential novels was Daniel Defoe's *Robinson Crusoe*, detailing the travails of a man marooned on a desert island. At the tail end of that century, English poet Samuel Taylor Coleridge published his famed ballad *The Rime of the Ancyent Marinere* (*The Rime of the Ancient Mariner*), detailing the experiences of a sailor who has returned from a long sea voyage and containing the oft-repeated lines:

Water, water, every where,
And all the boards did shrink;
Water, water, every where,
Nor any drop to drink.

In North America in the mid-1800s, Herman Melville was busy writing several books focused on the sea, surely the most famous of which is *Moby-Dick*. In this masterpiece, Sailor Ishmael relates the tale of the obsessive quest of Captain Ahab for revenge on Moby Dick, the great white whale that previously bit off Ahab's leg at the knee. Shortly thereafter, the first modern American novel stemmed from the pen of Mark Twain, once again centered on water as a theme. *The Adventures of Huckleberry Finn* shares the story of a boy and a runaway slave as they raft down the Mississippi River. Henry David Thoreau's iconic 1854 journal *Walden* is wrapped in connections to a pond. Across the Atlantic in France in 1870, Jules Verne published his classic science fiction adventure novel *Twenty Thousand Leagues Under the Sea*. In this work, he describes a fantastical underwater ship — the Nautilus — captained by Nemo. Features of this vessel accurately anticipated technologies for submarines that only appeared decades later. Building on the inspiration of *Walden*, contemporary writers in the middle of the 20th century became increasingly interested in the health of the environment. Among the most influential productions was Rachel Carson's 1962 book *Silent Spring*, which documented the adverse effects on water of the indiscriminate use of synthetic pesticides like DDT. This book is widely credited as launching the modern environmental movement that led to the creation of the U.S. Environmental Protection Agency.

Water also has a central presence on the big screen, often representing a hostile environment in films. Few people of a certain generation can venture into the water at the beach without at least a momentary hesitation thanks to the 1975 blockbuster *Jaws*. *Jaws* provided the solution to insurmountable challenges: "We're gonna need a bigger boat." The water itself takes on an ominous role in *Deliverance*, provides the ubiquitous backdrop for a post-apocalyptic future in *Waterworld*, and seems like an alien world in *The Abyss*. Who could forget the obviously water-based 2003 hit produced by Pixar *Finding Nemo* (save, perhaps, Dory, Nemo's chronically memory-challenged regal blue tang sidekick)? Merging literature and film, *The Little Mermaid* recast as a Disney animated musical fantasy film Hans Christian Andersen's 1837 fairy tale, telling the story of a beautiful mermaid princess who dreams of becoming human. More recently, Guillermo del Toro's *The Shape of Water* introduces the audience to a mysterious amphibian captured from a South American river, in some sense recasting the story of the 1954 classic horror film — also featuring water — *Creature from the Black Lagoon*, but with a less violent (and more romantic) outcome.

Not surprisingly, water has appeared in artistic creations since the earliest days of human civilization. While water is perhaps the essence of stillness and serenity, it is water's characteristic movement that often shows up in art. The Minoans of Crete typically stylized water in curvilinear forms such as spirals, whereas the ancient Egyptians chose the zigzag in their hieroglyphs (Figure 7.1). The 11th-century Bayeux Tapestry opted for embroidered wavy lines in its representation of the English Channel.

Leonardo da Vinci, the virtuoso Renaissance inventor and artist, was captivated by water. da Vinci sketched water in great detail and drew on his observations for his designs of water transportation systems. Elsewhere in Europe, Dutch artists established a tradition for marine art, which spread to Britain and France. In the latter, Claude

spirals (Minoan)

zigzags (Egyptian)

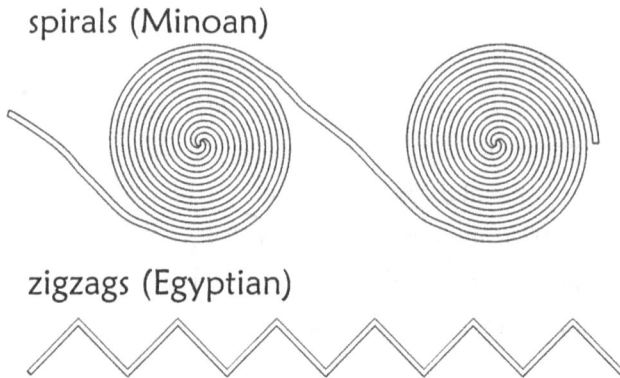

Figure 7.1. Stylized forms used to represent water in ancient art.

Monet painted numerous seascapes, and his 1874 oil sketch of the Le Havre harbor *Impression: Sunrise* lent its name to the Impressionist movement. America, too, produced prolific landscape painters who relied heavily on water scenes such as those from the 19th century Hudson River School. In the Far East around the same time, Japanese artist Hokusai created a woodblock print called *The Great Wave off Kanagawa*, also known as *The Great Wave* or simply *The Wave*. This work, probably the world's most recognizable piece of Japanese art, depicts a mammoth wave threatening boats off the coast of Kanagawa. While coastal residents may go generations between experiencing a tsunami, *The Great Wave*, keeps the risk on our minds. One can find water-inspired works throughout folk art traditions as well, from carved fish decoys to ship's figureheads to scrimshaws. Vietnamese folk artists have long performed shows using water puppets, in which the puppeteers stand in water behind a screen with the bamboo puppet theater floating in a pond. In modern times, the Canadian performance company, Cirque du Soleil, has "weaved an aquatic tapestry" into their water-inspired show, *O* in Las Vegas since 1998. Architecture, too, incorporates water themes, perhaps most notably in Frank Lloyd Wright's brilliant creation Fallingwater.

While the sounds of water's movement provide a natural backdrop at Fallingwater, musicians through the ages have both drawn inspiration from water and weaved water themes into their sounds. Classical composers often drew on water as their muse. Claude Debussy's 1905 orchestral composition *La Mer* (*The Ocean*) transformed the softness of impressionist art into harmonies that bring to mind images of waves and droplets of their spray pattering on the surface. Debussy's younger contemporary, Maurice Ravel, shared his impressionist style and he, too, created water-inspired pieces such as the piano suite *Miroirs*, which contains the movement *Une barque sur l'océan* (A Boat on the Ocean). This section recounts a boat as it sails upon the waves, with sweeping melodies emulating the flow of ocean currents. While it actually isn't about water *per se*, we would be remiss if we didn't also mention George Frideric Handel's collection of orchestral movements, *Water Music*. In this case, the name derives from the location of its inaugural 1717 performance: the River Thames. Massive barges housed the musicians, who played for none other than King George I.

Songs that incorporate lyrics offer the opportunity to convey messages with more than just melodies. The Great Mississippi Flood of 1927 was the most destructive river flood in the history of the U.S., inundating 27,000 square miles with water up to 30 feet deep. The devastating impacts of this flood found their way into a number of songs, ranging from Blind Lemon Jefferson's *Rising High Water Blues*:

Backwater rising, Southern peoples can't make no time
I said, backwater rising, Southern peoples can't make no time
And I can't get no hearing from that Memphis girl of mine.

to another blues track, written by Kansas Joe McCoy and Memphis Minnie and famously remade in the 1970s by rock group Led Zeppelin: *When the Levee Breaks*. Flipping to the challenge of insufficient access to water, one of the top Country Western songs of all time is *Cool Water*, written by Bob Nolan in 1936, is about a man and his mule and a tantalizing mirage leading them to erroneously believe relief was on the horizon. Water's centrality to sustaining — or ending — life is also captured in Fela Kuti's *Water No Get Enemy*, mixing vocals of the Yoruba language and pidgin English from West Africa to highlight the power of water in both literal and metaphorical senses. Fully metaphorical uses are equally abundant in music, such as the reggae standard *Till Your Well Runs Dry* by Peter Tosh, in which he assures his recent ex that she will miss him when he's gone. The tranquility of water pervades classics such as Otis Redding's *(Sittin' On) The Dock of the Bay*, inspired by his stay at a boathouse in Sausalito, California on the 'Frisco Bay, and Henry Mancini's *Moon River*, the latter sung by Audrey Hepburn in the 1961 film *Breakfast at Tiffany's*. The lyrics, written by Johnny Mercer, are evocative of his youth spent in Savannah, Georgia, and its lazy waterways. Water can also bring straight up fun to a song. Pretty much the entire library of The Beach Boys tracks falls in this category, as does The Band's *Up on Cripple Creek*:

> When I get off of this mountain
> You know where I want to go?
> Straight down the Mississippi River
> To the Gulf of Mexico
> To Lake Charles, Louisiana
> Little Bessie, girl that I once knew
> She told me just to come on by
> If there's anything she could do
>
> Up on Cripple Creek, she sends me
> If I spring a leak, she mends me
> I don't have to speak, she defends me
> A drunkard's dream if I ever did see one.

There are plenty of examples with water in its familiar, solid form as well. *Ice Ice Baby*.

Water in religion

Water and religion are seemingly inseparable. Water cleanses, whether it be objects for ritual use or people preparing both physically and spiritually to worship. We are at water's mercy, depending on it every day of our lives, as many believe we are at the mercy of God or gods. Judeo-Christian religions integrate water in many ways. Stories from the Torah such as the Great Flood in Genesis and the parting of the Red Sea during the Exodus show water's use as a powerful instrument by God to deliver punishment. Indeed, such power over nature on a grand scale is a compelling illustration of God's supremacy. Judaism is rife with washing rituals, from *mikveh* baths to the cleansing of hands or feet. Baptism is central to much of Christendom, with origins both in the biblical story of Moses and the parting of the Red Sea and in the baptism of Jesus by John the Baptist in the Jordan River. Holy water, blessed for use in rites, is applied in dedications, burials, blessings, and a number of other occasions. Christians also speak of "living water," referencing the story of Jesus offering living water (meaning himself) to a Samaritan woman so that she will never thirst.

Islam makes heavy use of water for cleansing. Mosques often have pools of clear water so that Muslims can be ritually pure before coming to prayer. Fountains, too, are a common feature of Muslim places of worship. Islamic ablutions include *ghusl*, the major ablution, entailing the washing of the entire body; *wudu*, the minor ablution, generally including washing of the face, head, hands up to the elbows, and feet to the ankles; and a third type in which water is not available, sometimes substituted by clean sand. Buddhism, in contrast, often shuns ritual and symbolism, but in Buddhist funerals, water is poured into a bowl to the point of spilling over its edges, with monks reciting "As the rains fill the rivers and overflow into the ocean, so likewise may what is given here reach the departed." Shintoism, native to Japan, entails worship that must be preceded by purification with water, with shrines often housing troughs for this ritual. Waterfalls, in particular, are considered sacred in this faith.

Hindus believe all water is sacred, especially rivers, and particularly the seven sacred rivers (Ganges, Godavari, Kaveri, Narmada, Sarasvati, Sindhu, and Yamuna). The Ganges is most important among these. Its waters are said to flow from the toe of Vishnu, and those who bathe in the river and leave some part of themselves on the left bank — such as some hair — will attain the paradise of *Svarga*. Holy places in the Hindu faith are often situated on the shores of water bodies, and especially at the convergence of several rivers. Funerals are held near rivers, and after cremation in a pyre, the ashes are spread in the river. Hindu scripture also has a flood story, the Great Flood of Manu, with Manu (similar to Noah) being the sole survivor of a worldwide deluge. Zoroastrians also believe water to be sacred, and they believe strongly that it must not be polluted. In this tradition, when the world was created, the Evil Spirit Angra Mainyu poisoned the pure water ocean with salt. Ablution is a central practice to cleanse away pollution. And yes, Zoroastrianism also has a flood story. Yima was instructed by Ahura Mazda to construct a massive boat to house specimens from each plant and animal species prior to a great flood. It seems there is some universality to the belief that water can cleanse not only the individual, but an entire planet if necessary.

Water in recreation

Beyond such serious matters, water is almost a universal implement for recreation. In virtually all cultures, our recreation on hot days involves swimming, whether in pools, lakes, or oceans. Beaches are a primary destination for vacations. People will toil throughout the year to spend a few days sitting near the ocean and listening to the waves and smelling the sea salt. Water parks, with their slides, flumes, and wave pools, draw thousands every year in search of some refreshing recreation. At home, nearly any outdoor surface can be transformed into a low-friction sheet of fun with a slip-and-slide, and water gun and balloon fights have entertained generations of children. There is an entire sports and activity culture focused on water including swimming, diving, surfing, water polo, snorkeling, scuba diving, surfing, water skiing, boating, and fishing. A major portion of the summer Olympics is focused on water activities. Just as dominant, frozen water, whether as snow or ice is the choice for winter recreation including downhill skiing, cross-country skiing, sledding, ice skating, hockey, and even ice fishing. The winter Olympics would not exist without frozen water. Imagine a world without curling, if you can, or Jamaican bobsledding.

08 Water is the Future

A discussion of the future of water must begin with a look at how our changing climate is already impacting water, and how these effects will be substantially amplified in coming years. Regardless of the success of global efforts to mitigate climate disruption by reducing carbon emissions in the ongoing shift from fossil fuels to renewable energy sources, further changes to the climate are inevitable in the short-to-medium term. While energy is the primary agent driving climate change, water is the central medium through which changes in our climate affect our ecosystem and, thereby, the well-being of people. Like water, climate does not respect borders. Adapting to climate-induced water stresses will necessitate cooperation at community, national, and regional levels.

Effects of climate disruption on water supply

Climate and water are intertwined in innumerable aspects. A central connection is the influence of climate on the hydrological cycle, impacting water supplies in a big way. The rate of water evaporation from a surface depends on many factors, including water temperature, air temperature, air humidity, and air speed above the water surface. As global temperatures continue to climb relentlessly, driven by enhanced greenhouse warming, evaporation of water from bodies of water and soils accelerates. Plants, too, join the party with increased evapotranspiration. All that extra water entering the air has to go somewhere. The atmosphere can accommodate additional moisture when it is warmer—about 4% more for every degree Fahrenheit rise (or about 7% for 1° C rise)—so a share of the additional moisture will increase moisture levels in the lower levels of the atmosphere (troposphere). The rest will fall as precipitation, eventually returning to the soil and bodies of water as part of the endless cycle. While the overall quantity of precipitation is likely to increase somewhat, the frequency, intensity, distribution, and form of that precipitation are all projected to change dramatically.

We don't have a good way to measure how much precipitation falls on the oceans, which of course represent the majority of the surface of our planet. That said, measurements of overall precipitation on land do provide a reasonable proxy for precipitation in general. It is important to keep in mind that, when discussing climate change, only trends over a period of several decades can be considered significant. Weather varies from season to season and year to year for lots of reasons, but when average changes display trends over 20 or 30 years, they most likely represent a true shift in the climate. Warming of our planet has already resulted in measurably increased precipitation. On average, total annual precipitation has intensified over land areas in both the United States and worldwide (Figures 8.1 and 8.2). Since the beginning of the 20th century,

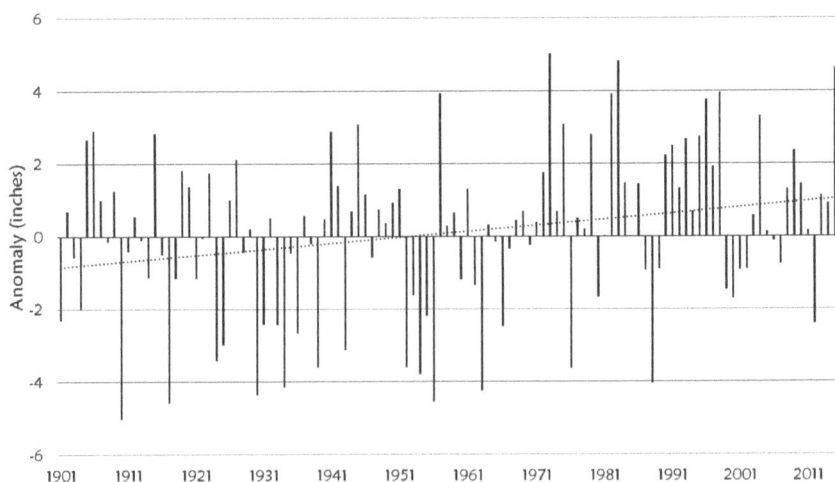

Figure 8.1.   Average annual precipitation in the United States from 1901–2015. Data from NOAA.

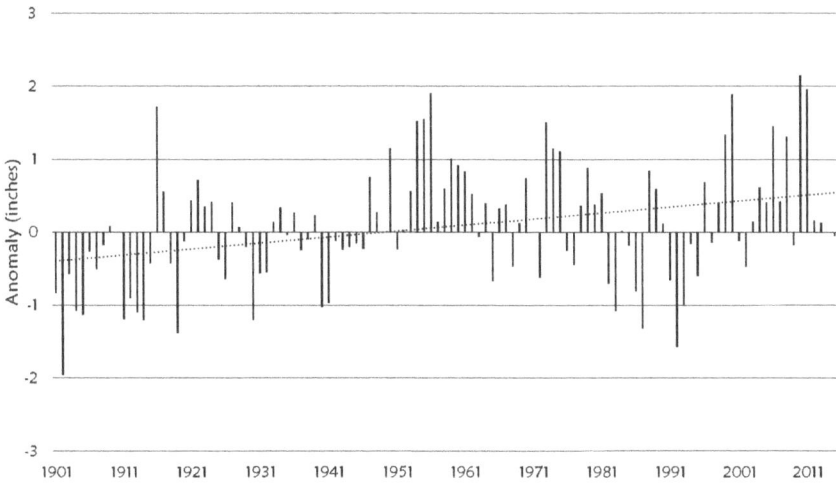

Figure 8.2.   Average annual global precipitation from 1901–2015. Data from *State of the Climate in 2013* (DOI: http://dx.doi.org/10.1175/2014BAMSStateofthe Climate.1) with 2016 web updates.

roughly coinciding with large increases in industrial activity and fossil fuel consumption, global precipitation has risen at an average rate of 0.08 inches per decade. Demonstrating that such changes are not evenly distributed around the world, precipitation in the contiguous 48 states has increased at more than twice this rate: 0.17 inches per decade. Obviously, that means other parts of the world have received less precipitation on average over the same period.

There is also substantial variability in precipitation trends within a given country, especially within countries with large geographic area. In the United States, heavy precipitation events are already increasing in the Midwest, Northern Great Plains, and Northeast; in contrast, the Western U.S. — and the Southwest in particular — is seeing less overall precipitation. The fact that the increased precipitation is concentrated in especially strong storms means that more of the water will rapidly runoff into waterways, translating into destructive soil erosion, less recharge of groundwater aquifers, and rampant flooding. Another consequence of high-intensity runoff is that hydropower production tends to suffer since reservoirs often have to be managed with excess spillover to avoid overfilling during extreme precipitation events. Even in regions that experience less overall precipitation, the rain that does fall will tend to arrive in concentrated storms, further straining the water supply. This change in rainfall patterns is often discussed in terms of "IDF," or intensity, duration, and frequency. With increasing IDF, neither natural systems nor human-designed infrastructure is well adapted to rainfall, potentially resulting in flooding, damage, and mudslides.

Like the American west, regions such as the Mediterranean, southern Africa, and Australia are also projected to experience less overall precipitation in the coming decades. Droughts are inevitable. In 1965, a meteorologist working for the U.S. Weather Bureau named Wayne Palmer developed an index to quantify droughts known today as the Palmer Drought Severity Index (PDSI). It is a measurement of dryness, based on precipitation and temperature data, which models soil moisture based on supply and demand of water. Supply data are relatively simple — that's just the amount of precipitation in a given area. Demand, however, is challenging to quantify since it is affected by temperature, preexisting moisture, soil type, rate of evapotranspiration, and other factors. The PDSI is an attempt to overcome this challenge by statistical correlations with temperature alone for the demand projection. While PDSI is certainly not a perfect index in projecting dryness, it has proven quite effective in correlating with long-term drought. Figure 8.3 depicts the results from a statistical analysis of the PDSI averaged over the globe based on data from the entire 20th century. While the data exhibit a lot of noise as a result of various shorter-term phenomena (such as the El Niño Southern Oscillation and other oceanic cycles), the long-term trend is clear: drought severity has increased worldwide. It is probable that agricultural activity will need to progressively shift to more northern latitudes, tracking the rainfall, to adapt to these climatic changes in precipitation (and temperature).

Beyond changes to overall levels of precipitation and extreme precipitation events, warmer temperatures will also have direct impact on the form of precipitation, namely, by producing a greater proportion of rain relative to snow. This change, though seemingly innocuous,

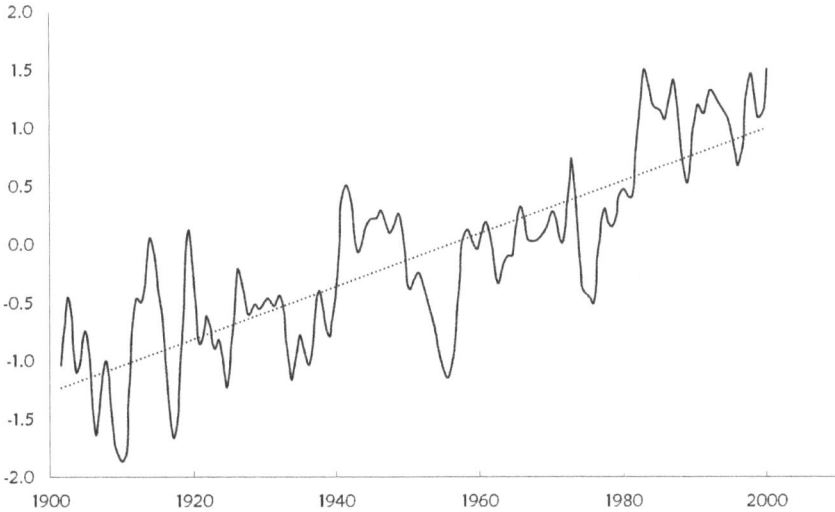

Figure 8.3.   Primary result from principal component analysis of global Palmer Drought Severity Index (PDSI) data showing an increasing trend in drought over the 20th century. Data from National Center for Atmospheric Research.

can have disastrous consequences. Snowfall in colder months acts as a form of water storage that is subsequently dispensed gradually over the course of the spring season as the snow slowly melts. When more of that snow falls as rain, it cannot be stored and rather runs off quickly, providing increased flow during the winter months when demand for that water is at its lowest levels of the year. Animals (including people) and plants need more water when it is warmer because of increased transpiration, and that is, of course, not during the winter. Moreover, the spring melting is occurring earlier and earlier in the year as average temperatures climb, once again shifting the water supply to a time when it is less valuable. In many regions of the world, such as the Sierra Nevada Mountains in Northern California, snowpack runoff is the primary source for river flow.

Another connection between a warming planet and water runoff is in the melting of glaciers. Glaciers are found on all continents except Australia. Glacier ice volume is governed by the balance between accumulation of snow, which compacts into ice, and melting or calving.

Climate change is altering this balance. Historical photos of glaciers, such as those collected through the Extreme Ice Survey, provide shocking visuals of how rapidly and significantly glaciers are disappearing across the globe. Beginning in the 1970s, a new tool emerged for tracking this process: satellites. Orbiting satellites provide compelling imaging, and they can also track gravitational changes associated with the loss of ice from glaciers and ice sheets. The satellite era paints a picture of accelerating ice changes in places as geographically dispersed as Alaska, Antarctica, and Greenland. Ice making up glaciers often takes centuries to develop, yet it can vanish in just a few years. Glaciers don't generally melt slowly and steadily. Once glacial ice begins breaking down, the meltwater and intruding seawater interact with the glacier, causing increasingly rapid melting and retreat. Initially, glacial melting will provide some extra water supply via runoff. As the ice diminishes, however, that runoff will diminish. Considering that meltwater runoff from glaciers is the primary source of water for billions of people on Earth, this is an alarming outlook. An additional challenge is that, because revegetation of terrain happens more slowly at high altitudes, deglaciated mountainous areas will experience increased erosion, threatening infrastructure.

Glaciers are not the only frozen things melting on Earth due to planetary warming. Permafrost is as well. Permafrost is a layer of rock and soil, located in countries at high latitudes, which usually is frozen throughout the year. As surface temperatures rise, the water within the permafrost thaws. Because the mechanical properties of the soil change substantially upon thawing, the land sitting above the permafrost layer changes shape or even sinks. This sort of subsidence can damage buildings and infrastructure — including (or especially) water and sewer pipes, which threatens distribution and collection systems. Reshaping of the natural landscape also impacts how water flows along the surface, rerouting traditional paths and shifting drainage patterns. Moreover, subsurface water is mobilized when it is released from the previously frozen permafrost. Both of these changes to local hydrology directly impact ecosystems.

In addition to these effects on the hydrological cycle, there are further ways in which a changing climate will impact water supplies. Akin to the manner in which permafrost melting will inflict damage to water delivery infrastructure in northern regions, coastal region water infrastructure will face increasing threats from storm surges. When a low-pressure weather system such as a tropical cyclone approaches land, the water level rises beyond that produced by typical astronomical tides. This storm surge is produced by water being pushed toward the shore by the winds moving cyclonically around the storm's center. Climate change is projected to increase the power of major storms and, when coupled with (also climate-change-exacerbated) degradation of coastal wetlands that have historically shielded against such events, amplified storm surge devastation is inevitable. Water purification and treatment facilities, reservoirs, pumping stations, and pipe networks are all potential targets for these surges.

A more tangential but no less dangerous connection between disruption of the climate and water is the increasing spread of water-related vector-borne diseases. This class of infections is transmitted by the bite of infected mosquitoes and other arthropods that rely on water for pivotal stages of development. Shifting climatic conditions lead to these species reaching into geographic locations where they previously did not have a foothold. Chikungunya, first identified in Tanzania in the 1950s, is a virus that attacks joints. The name originates from the Kimakonde language and translates roughly as "to become contorted," which is the effect upon humans unfortunate enough to suffer from severe symptoms. Chikungunya has plagued Africa and Asia for decades, but in recent years, it is gaining traction in places such as Italy. Similarly, dengue fever has recently spread in regions as far separated as Argentina, Australia, and China. Ailments such as malaria and West Nile virus are also dispersing to new parts of the globe, and the mounting danger from the previously underestimated Zika virus has grabbed headlines in recent years with the realization that it can lead to microcephaly, severe brain malformations, and other birth defects.

Effects of climate disruption on water quality

Climate disruption is not only affecting the amount, frequency, intensity, distribution, and form of precipitation, but it is also directly connected to the quality of the water available for human use and for the ecosystem in general. While pollutants dissolved or suspended in the water are probably the first thing that comes to mind regarding water quality, and indeed a critical subject that we will come to shortly, there is a more subtle yet no less important threat to water quality: heat. As global temperatures creep ever higher, the average temperature of water bodies climbs along with them. When water temperature varies, so does the concentration of oxygen dissolved in that water. Cold water can maintain more dissolved oxygen than warm water. Dissolved oxygen in surface water is used by essentially all forms of aquatic life. Oxygen enters water from the atmosphere (and, to some extent, from groundwater discharge). Dissolved-oxygen concentrations fluctuate with water temperature both on short time scales, such as over the course of a day, and on longer time scales, such as with seasonal shifts. Global warming, however, is gradually increasing the overall average (Figure 8.4).

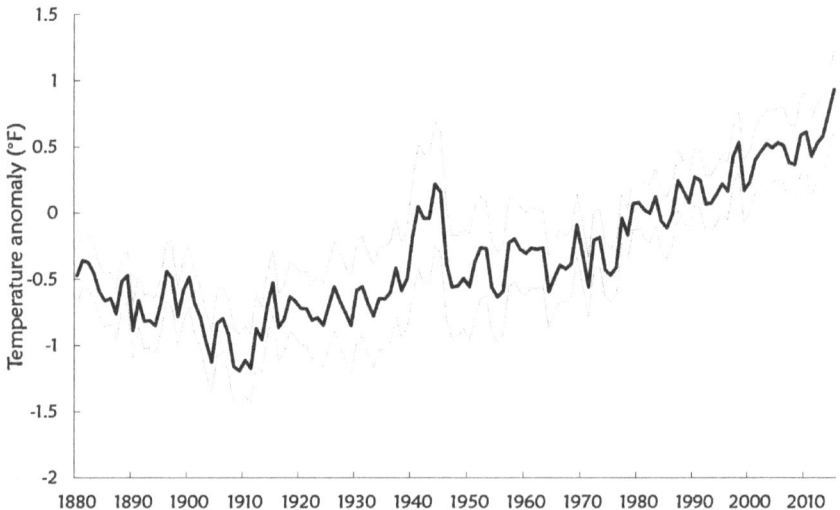

Figure 8.4. Average surface temperature of the world's oceans since 1880, using the 1971 to 2000 average as a baseline for depicting change. The dashed lines show the range of uncertainty in the data. Data from NOAA (2016).

The resulting decrease in oxygen concentration in the water stresses fish, insects, crustaceans, and many other organisms. Warmer water also exacerbates algal blooms by accelerating the algae's rate of reproduction. In the summer, toxic algal blooms (TABs) can develop rapidly and impact public drinking water supplies. In the United States, southern Lake Erie is at highest risk of a major disruption of drinking water. The TABs happen because of a confluence of several factors: nutrient runoff from agriculture, shallow water, and inlets with slow exchange with the bulk of the lake water. Human activity, too, suffers from warmer water. As discussed earlier in this book, electricity production represents the largest water withdrawal sector in the United States, and a major sector for water use worldwide. The water is used to cool power plants, as well as to generate the steam to turn turbines for electricity generation. When the temperature of the cooling water is higher, the efficiency of power production decreases. On numerous occasions in recent years, plants have even had to be derated (reduce power output) or shut down because water temperatures were too high. In the summer of 2012, the Millstone nuclear plant in Waterford, CT shut down because the water in Long Island Sound was too warm to cool it. The same thing happened in 2015 at the Pilgrim Nuclear Power Station on Cape Cod Bay. Similar events have occurred in states as geographically disparate as Pennsylvania, South Carolina, and Illinois, as well as in Canada, Europe, and beyond. It has been projected that thermoelectric power generating capacity will decrease by 4–16% in the U.S. and 6–19% in Europe by 2060 due to lack of cooling water.

Another connection between temperature and surface water is an effect known as thermal stratification. In this case, however, the water quality concern is not a direct result of rising temperatures, rather, it is an effect of decreasing flow of water in shallow lakes. Thermal stratification of lakes refers to the fact that cold water is denser than warm water, and so it tends to sink to lower depths. In the absence of any forces driving vertical mixing, a lake will exhibit layers in which colder, denser water sits at the bottom. In temperate regions of the planet, lake water varies in temperature through the seasons, and this cyclical pattern produces overturn as the water near the surface cools in winter and sinks. Another cause of mixing is the natural inflow and outflow of water to and from rivers. If there is insufficient mixing and the thermal stratification of water persists for long periods, the lake is said to be "meromictic." Many regions forecast to become drier in the future will see more meromictic lakes. These lakes can succumb to large fish die-offs as a result of accumulated dissolved carbon dioxide. Nutrient and pollutant loads also increase when seasonal mixing decreases, leading to blooms of toxic species and degradation of the quality of water often relied upon for drinking.

Both drier and wetter regions will experience an increase in heavy precipitation events. When a downpour occurs, it can overwhelm municipal treatment systems. This is especially a problem in cities with combined sewer systems (CSSs), where the stormwater and sewer pipes are shared. There are hundreds of such systems in the United States alone, clustered primarily in the Northeast, Midwest, and Northwest states. Most CSSs were installed in older cities and municipal regions. They were designed without considering climate change, increased runoff from impervious surfaces, or increased residential water use. Under normal conditions, a CSS transports wastewater to a sewage plant for treatment, then discharges to a river or lake. During heavy rainfall, the volume of wastewater can exceed the capacity of the system. When this occurs, untreated water discharges directly to water bodies or backs up into residences and onto streets. Effects on water quality are horrific. Beyond combined sewer overflows, deluges of rain dramatically increase runoff regardless of the design of a local sewer system. Runoff carries a mélange of undesirable materials into water supplies. Nutrients from agricultural plots or fertilized lawns, sediment, industrial pollutants, animal waste, and untold other contaminants, always a challenge at some level, become serious public health threats during these events.

One inevitable consequence of a warming planet is a rise in sea level (Figure 8.5). There are two mechanisms contributing to sea level rise: (1) as heat content of the oceans increases, the water itself expands and (2) melting of principal stores of land ice such as glaciers and ice sheets adds water to the oceans. (Note that sea ice melting has no direct net effect

Figure 8.5.   Changes in sea level for the world's oceans since 1880, using long-term tide gauge measurements (solid line) and recent satellite measurements (dotted line). Average absolute sea level change is plotted, which is independent of whether nearby land is rising or falling. The dashed lines show the uncertainty in these data. Data from CSIRO (2015) and NOAA (2016).

on sea level, just as ice cubes melting in your beverage do not change the level in your glass. It does, however, accelerate warming by reducing the reflectivity of the sea surface.)

Rising seas increase vulnerability to erosion and flooding along coasts, further aggravating the threat from the storm surges discussed previously. Regarding water quality, though, the primary concern is saltwater intrusion into freshwater sources. Intrusion of ocean water into fresh rivers is already happening as withdrawals from those rivers for human use increases, but sea level rise will inexorably push ever more salty water upstream. The Biscayne aquifer, covering about 4,000 square miles in south Florida, is recharged from the Everglades, which are becoming increasingly saline. Further north, aquifers in New Jersey are recharged by the Delaware River, which is also becoming increasingly saline, particularly during droughts. Fresh coastal aquifers are also at risk

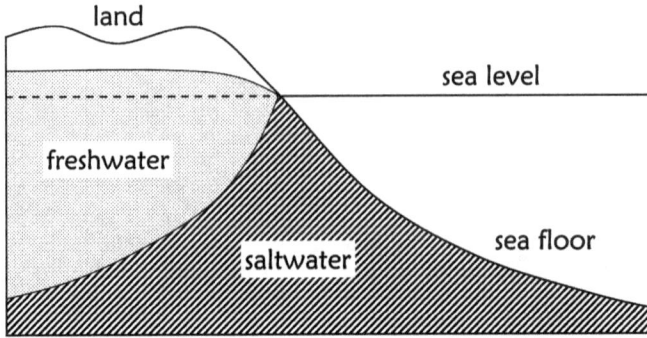

Figure 8.6.   Diagram of coastal margins where fresh groundwater meets seawater.

of saltwater intrusion. At coastal margins, fresh groundwater originating inland meets saline groundwater incurring from the ocean. This hydraulic connection naturally leads to saltwater intrusion. Saline water is denser and has a higher water pressure, so it can push inland beneath the freshwater in an aquifer (Figure 8.6). Pumping groundwater from coastal freshwater wells worsens saltwater intrusion because water extraction lowers the level of the fresh groundwater, reducing its water pressure and permitting salty water to flow further inland.

Sea level has already risen about nine inches since the start of the Industrial Revolution, and it will continue to rise for the foreseeable future — at an accelerating pace. Saltwater intrusion is already a challenge in countries in North and West Africa such as Tunisia, Morocco, and Benin, as well as in Asia and the Middle East (e.g. Pakistan and Cyprus). The United States is experiencing intrusion on all its coasts, from Florida and Louisiana to California and Washington. Higher seas will only make these problems graver and more widespread.

Beyond salt water intrusion, sea level rise is an existential threat for island nations in the south Pacific and the Indian Ocean. While nine inches may seem minimal, many island nations average only about one meter above sea level elevation. Sea level rise is expected to exceed one meter in the 21st century, essentially wiping out island homelands.

Mitigating climate disruption is a necessity for long-term human survival and prosperity, but based on the current climate scenario and a realistic timeframe for shifting global energy supplies to lower carbon emissions, some level of adaptation will be essential. One can look at this need as an opportunity. Adaptation to increased variability in climate will have a constructive side effect, namely, improved and more holistic water management overall. In the following sections, we will explore what the future holds for managing our water resources, including the prospect for technological and data innovations to open non-traditional water sources and to use existing sources with enhanced efficiency. The dire warnings of this chapter point to a new vision for water and society; if we act with urgency and care, the future can point toward sustainable, secure, safe, and affordable water supplies.

The future of water management, technology, and policy

The overarching theme of the future of our relationship with water is resilience. Unprecedented challenges are on the horizon, and we need to prepare ourselves to face them head on. Adaptation to an ever-evolving water scenario necessitates intimate cooperation. Sectors that have traditionally operated independently are interwoven by their connections with water. Energy, industry, land use, and — of course — water sectors must be viewed holistically in order to accommodate shifting supplies and demands. Because water pays no attention to borders, cooperation must also happen on a hydrological scale rather than a political one. Successful adaptation demands that, first, we maintain the existing water infrastructure we already have. This is no small task. Drinking water is carried by about a million miles of pipes across the United States. Much of that network was put in place in the first half of the 20th century, with a typical lifespan of 75–100 years. There are an estimated 240,000 water main breaks each year in the U.S., resulting in over two trillion gallons of treated drinking water being wasted. As a comparison, two trillion gallons of oil would last the U.S. almost seven years. The American Water Works Association projects that a $1 trillion investment — approximately 5% of all Federal non-defense discretionary spending — is needed to maintain and expand the service to meet the country's demands over the next 25 years. And that's just the United States. Europe's maintenance needs are estimated at more than $250 billion over the next decade. A similar story can be told around the world. The tools for finding and correcting leaks are improving. For example, companies market software tools that "listen" to the flow to identify and locate leaks.

Assuming the leaks and expanded service are addressed, management of future water resources is still a daunting challenge. Smarter administration of both water supply and demand are needed. Operation of water systems will need to accommodate a perpetually increasing array and scale of pollutants as well as major shifts in demographics (e.g. urbanization, relocation closer to resources, population growth). These changes are best addressed from the top down, presuming the cross-sector and political boundary cooperation discussed above is implemented. Demand management, however, is a more complex situation. Advantages can be gained both by shifting (spatially and temporally) and by reducing demand. If implemented with consumer awareness, water pricing is a powerful tool to guide demand, as discussed earlier in this book. Conservation practices can be spurred by municipalities, but grassroots cultural changes will be necessary for them to succeed. Working with school-age children is an especially promising way to drive acceptance of conservation practices.

Beyond the colossal necessity for maintenance of water infrastructure, there is also a need for major new investments to expand capacity and performance. These run the gamut from ecosystem restoration to reservoirs for water storage to levees for mitigation of flooding to irrigation systems for multiplication of food production to new and improved treatment infrastructure for converting difficult sources into fit-for-use water. Fit-for-use implies that we will only treat water to the quality needed for the specific application. For example, cooling water does not need to be drinking water quality. Beside "gray infrastructure" such as levees, "green infrastructure" considers the way that nature manages water. Green infrastructure approaches include replacing impervious surfaces with permeable materials to adsorb water or designing landscape with plants and fauna that are suitable for the expected water quality. Underlying this strategy is a massive need for innovation. We are in desperate need of new water technologies.

In some cases, the issue is more about getting an existing technology into more widespread use. Drip irrigation is a prime example. As discussed earlier in this book, shifting from traditional irrigation strategies to a plant-scale drip system can translate into remarkable reduction in water use for agriculture, which is — after all — the largest consumer of water in the world. Other, somewhat more developmental, agricultural innovations can also produce substantial water-use efficiency, such as in the biotechnology arena: for example, the designing of crops that reduce water loss from transpiration during stress. This innovation reduces crop failure from a heat wave or drought. Demand for food is escalating worldwide, and the vast majority of feasible arable land has already been exploited to this end. This means that the productivity of existing farmland must be increased to feed the world — likely requiring a doubling of production while at the same time using less freshwater. A formidable challenge. Meeting future food needs in a world with water scarcity will almost assuredly necessitate the implementation of new genetically engineered crops. One apparent area for enhancement is in drought and/ or salt tolerance, whereby plants are endowed with an improved capability to withstand short periods of low soil moisture or more salty soil. A less obvious, but potentially more impactful, adjustment would be to augment the ability of food crops to resist pests and pathogens. Most agricultural losses take place after the plants are fully grown, at which point the plant's integrated lifetime water needs have already been invested. Reducing losses associated with various common pests and pathogens will therefore dramatically increase the overall water-use efficiency. The key here is to decipher the molecular mechanisms that impart resistance in natural plants, and then to transfer the genes responsible for those mechanisms from one plant species to another. The overall food production system is ripe for efficiency improvements. Practices are quite distinct in the developed and developing world. In the developed world, we are very efficient at getting food from the seed to the refrigerator and very inefficient after that. In the developing world, we are inefficient at getting the food from seed to the home but very efficient once it is in the home. Therefore, innovation will be required along the supply chain.

Another central strategy to adapt to the future water scenario is water recycling. All water is, of course, recycled already. Almost all of the water distributed on the planet has been around for billions of years, and Mother Nature has been repeatedly cleaning it up as it passes through the water cycle. The challenge with the water cycle is that it can be quite slow. How long it takes, exactly, will depend on the path the water takes from point A to point B. The shortest stop in the water cycle is the atmosphere, where water tends to stick around for a few days to a few weeks. Once it returns to the ground as rain, it can flow down as surface water, which can take a month or two. If it falls instead as snow, it will sometimes stick around for a few months until it melts in the spring. (If that snow gets packed down into a glacier, the water will be trapped for as long as a century.) If the water manages to trickle down into aquifers, it can stay there anywhere from a few centuries to hundreds of thousands of years. Most water resides in the oceans, and can remain there for thousands of years. The longest stop on the journey through the water cycle is the water that ends up locked in ice shelves like those found in Antarctica. Once buried there, it can remain for a million years or more. In a world where water demand is growing and supply from the traditional water cycle is shrinking, it makes sense to explore ways to shortcut the water cycle — wastewater recycling is certainly near the top of that list. Also known as "reclaimed water," this water can be used several times before returning to the natural water cycle. Because wastewater treatment processes rarely produce water with quality rivaling that coming into homes, most of the uses of water reclamation are non-potable. These can range from toilet flushing to cooling water for power plants to hydraulic fracturing operations. Often, systems utilize a dual piping system to keep the recycled water clearly separate from the potable water. In the future, water reclamation will need to expand from a niche process to a standard fixture in our overall water management portfolio. Moreover, as increasingly polluted sources are exploited for recycling, there is a massive need for new technologies both for identifying pollutants and — critically — for separating them from the water efficiently. Pollutants that have been presented at only trace levels may start increasing in concentration as we recycle water.

Before one can tackle the challenge of cleaning up water to make it fit-for-purpose, one has to know what the heck is in there that needs to be removed. For the filthy soup entering a wastewater treatment plant, it is visibly obvious that the water is not clean. But pristine looking water can also contain any of a number of hazardous things invisible to the naked eye. You might think we have technology available to monitor for an array of such things in real time, but you would be wrong. Generally speaking, the way one identifies the composition of an aqueous (water-based) solution in practice is to extract a sample and perform tests on it in a laboratory setting using analytical chemistry and biology techniques. This process is slow and labor-intensive. That's great for employment numbers in the chemical and biological sciences, but not so great for efficient and effective monitoring of water quality. It can take hours to get a result, and only a few key contaminants are even measured. Ideally, there would be a real-time stream of data from sensors integrated into various points within the treatment facility detecting a vast array of materials of interest (microorganisms, heavy metals, agricultural nutrients, radionuclides, disinfectants, trace organics, and so on). To achieve this ideal, there is a pressing need for sensor technologies that are selective, reliable, and sensitive. Armed with the information produced by such a sensor array, attention can turn to how to remove those materials deemed hazardous to the intended end use of the water. One area of notable advancement is biological detection. Using polymer chain reaction (PCR), we can now detect microorganisms, including potentially toxic species, even if there are only a few microbial cells in the sample, and has become a routine test. In the past decade, a new technology has emerged, called metagenomics to understand the microbiome (entire microbe population) in the selected environment. With metagenomics we can detect the presence of potentially toxic microorganisms based on genes present, even if the sample contains a large range of species or the toxic microbe has never been isolated in the laboratory.

One intriguing trend in the wastewater industry is a shift from viewing the stream as "waste" to seeing it more as a potential valuable resource. Nutrients, such as ammonia, nitrates and phosphates, are a prime example. Phosphates originate from phosphate rocks, which are processed to produce fertilizer used in agriculture. Runoff from farms moves a large amount of these nutrients into waterways, and humans themselves also transport them into water by consuming food and then producing waste. Excess nutrients in wastewater act as food for algae and other contaminants causing eutrophication in lakes and oceans — approximately half of the lakes and reservoirs around the world suffer from this phenomenon. Adverse ecological impacts caused by eutrophication include biodiversity loss, water toxicity, and increased turbidity, translating to billions of dollars in economic losses every year. Phosphorous, in particular, is an important resource because it is not renewable in the same way ammonia and nitrates are (the latter using plentiful nitrogen from the air as a source for synthesis). While we can always produce more ammonia fertilizer, it is energy intensive. Ammonia is most often produced by reacting natural gas with atmospheric nitrogen. If it were possible to recover phosphorous from wastewater that would not only help prevent eutrophication, but it would also capture an indispensable material for reuse. Microalgae, it turns out, are rather efficient at removing nitrogen, phosphorous, and even heavy metals from water under the right conditions. These algae can then be harvested and converted into fertilizer or other products. There is ongoing research to engineer higher selectivity in algae strains to pull out particular materials.

In wastewater treatment plants, a different approach is sometimes used to reclaim the nutrients. The sludge from the wastewater plant is injected into an anaerobic digester. The anaerobic digester goes through two biological reactions in the absence of oxygen. In the first step, the sludge is digested by bacteria in the acetogenic phase to produce organic acids such as acetic acid (vinegar). In the second phase, the organic acids are digested in the methanogenic phase to produce methane (natural gas) and carbon dioxide in a mixture called biogas. Biogas is a renewable energy source that can be used to generate heat, electricity, or replace fossil natural gas. The nutrients partition into the biosolids at the bottom of the digester. Biosolids are used as a fertilizer replacement based on their nitrogen content. One interesting area of research seeks to accelerate the digestion rate and increase the relative energy content, i.e. methane, in the biogas. In nature, carbon dioxide reacts with minerals including calcium and magnesium and precipitations as carbonates. Researchers have been supplementing digesters with natural forms of these minerals and observed both increased digestion rates and higher energy biogas.

One of the most powerful ways of cleaning water is to use membranes. These are barriers that are semi-permeable, meaning that they allow certain things to pass through while blocking others. In most cases, this selectivity is achieved by size exclusion — the membrane has pores of a given size that prevent materials larger than the pores from passing through, not unlike a colander used to drain your pasta or a drip coffee filter. Contaminants in water span many orders of magnitude, from macroscopic bits of stuff like hair and sand down to individual ions and molecules (Figure 8.7).

Membranes are categorized by the range of sizes of things they can filter out. Big stuff is handled by particle filtration. Microfiltration (MF) addresses things from a few micrometers to a few hundred nanometers, followed by ultrafiltration (UF) for materials spanning down to a few nanometers and nanofiltration (NF) for those around a single nanometer — the scale of many molecules. Removal of very small molecules or ions like the dissolved salts in seawater requires a

coal particle

sand

hair   particle filtration

pollen

yeast

atmospheric
dust

red blood   microfiltration
cell

bacterium

asbestos   virus   ultrafiltration

pyrogen   nanofiltration

Na⁺ Cl⁻   reverse osmosis
ions

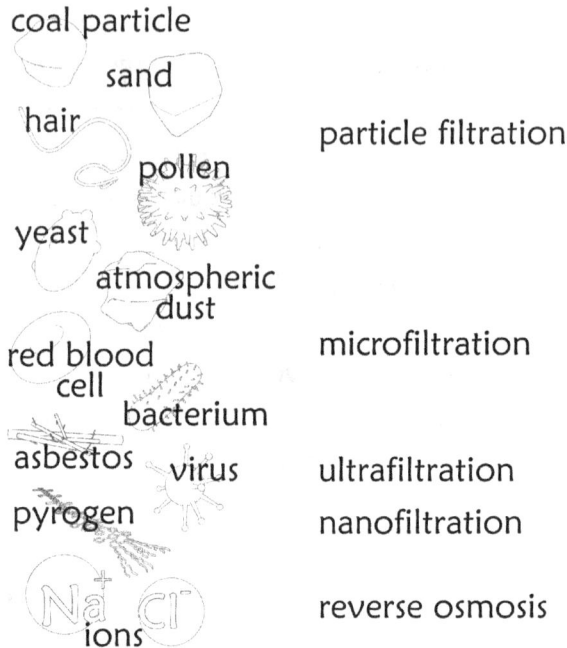

Figure 8.7.   Length scales of common pollutants in water and the corresponding membrane regimes used to remove them.

somewhat different mechanism, which is achieved using a process called "reverse osmosis" (RO).

Before you can understand reverse osmosis, you need to understand plain, old, vanilla osmosis. Osmosis entails spontaneous movement through a semi-permeable membrane of solvent molecules (e.g. water) from an area of low solute concentration into a solution of higher solute concentration. This process tends to make the solute concentrations on the two sides of the membrane equal, at which point the osmotic flow will balance and net flow will cease. Reverse osmosis, then, is the opposite — movement of solvent molecules to the membrane side with *lower* solute concentration, leaving concentrated gunk behind and producing clean water. To battle against the natural osmotic pressure in this way, you have to apply pressure to the feed solution. In typical membrane filtration, the process is governed by size exclusion. RO, in contrast, involves both size exclusion and diffusion. RO membranes are not even porous, in the traditional sense. Rather, water molecules jump around, slipping occasionally in between the polymer chains that make up the membrane. RO can never, therefore, remove 100% of certain tiny contaminants since they, too, can sometimes diffuse through the membrane. Regardless of the size regime in question, the need to apply pressure to squeeze water through a membrane translates to an unavoidable energy cost, and the smaller the pores, the greater that cost becomes. Despite the high energy demand for RO water treatment, this technology has blossomed in recent years, now representing the technique of choice for seawater desalination as well as for some wastewater treatment systems.

Membranes can take a number of different forms (Figure 8.8). A simple configuration is a big, flat sheet. This is the base unit for various other configurations, such as plate-and-frame (with a series of stacked flat

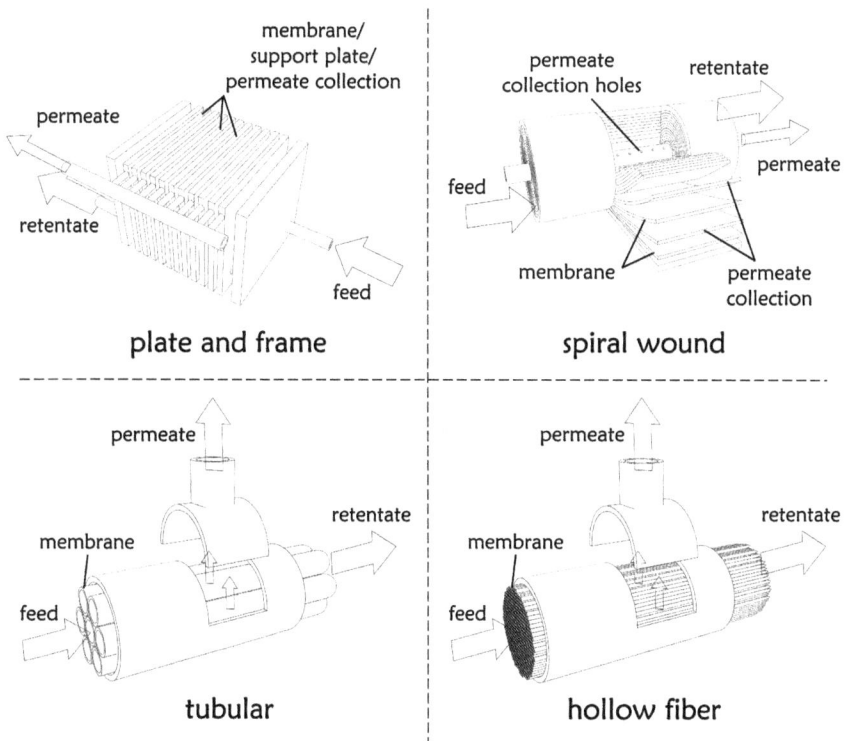

Figure 8.8.   Common configurations for membranes.

sheets and supporting structures) or spiral wound systems (with flexible feed and permeate spacers positioned between two flat sheets and the whole assembly rolled up like a jelly roll). Tubular systems, in contrast, involve placing membranes inside support tubes, which themselves are housed within a cylindrical shell to form a membrane module. Hollow fiber membranes, another common implementation, also involve a cylindrical housing, and they consist of a large number of small, hollow fibers where the feed stream can either pass from the cores to the outside of the fibers or vice versa. In all configurations, the nomenclature for the various streams involved are: "feed" for the incoming stream that requires treatment, "permeate" for the clean(er) water coming out, and "retentate" for the concentrated gunk that was removed from the feed by the membranes.

Generally, a membrane is considered effective if it removes a large fraction of the targeted material while allowing a maximum amount of water to pass through with minimal energy expenditure. A key parameter, therefore, is the membrane's "permeability," which is a measure of the membrane's ability to allow fluids to pass through it. Permeability is related to overall porosity (the fraction of the membrane's total volume that is made up of pores, or voids) as well as the size/shapes of the pores, how they are interconnected, and the chemistry of the pores' sidewalls. When operating in a size-exclusion regime, it is also important that there not be an excessive number of interconnected large pores in the membrane through which undesirable materials might find a way to pass. Today's commercial membranes do not always live up to this challenge, meaning that some amount of the material one wants to remove from the water remains in the permeate. In an ideal scenario, a membrane would have all of its pores exactly the same size such that all targeted materials would be trapped and end up in the retentate. Achieving this so-called isoporosity, however, is a daunting challenge in practice — especially if the cost of the membranes is to remain reasonably low. Cutting-edge research is exploring ways to scalably create membranes with narrow pore-size distributions; membranes that are even, perhaps, ultimately isoporous. Effectiveness of membrane separations would skyrocket if scientists and engineers manage to pull off this feat.

Most commercial membranes today are made up of polymeric materials (e.g. polysulfone, polyvinylidene fluoride, and polypropylene). While such materials are very cheap, they suffer from relatively poor stability when operating at elevated temperatures or under harsh chemical environments, and this limits their utility in real-world applications. Elevated temperatures increase the mobility of water molecules and increase permeability.

Recently, ceramic membranes have gained a foothold in the market for this reason. Ceramic membranes are made from inorganic materials such as alumina, titania, and zirconia oxides. These inorganic materials can stand up to aggressive chemical environments, such as highly acidic solutions, and exhibit excellent thermal stability — but these benefits carry concomitant increased costs. All membranes, polymeric or ceramic, face another pervasive challenge: fouling and scaling. Fouling is the accumulation of unwanted material from the feed stream on membrane surfaces, which can consist of either living organisms, known as biofouling, or non-living substances such as proteins and miscellaneous other organic molecules. Precipitation of inorganic chemicals on the membrane is called scaling. Both of these phenomena impede flow of water through the membrane by narrowing or blocking pores, which means that a larger pressure much be applied to maintain the desired flux, increasing the energy consumed by pumps. Eventually, the membranes must be cleaned (physically or chemically) or replaced. An exciting future direction in membrane research is to design membranes that can resist attachment of foulants and/or scalants, prolonging their useful lifetimes. Even more enticing is the prospect of membranes that can clean themselves. How could such a thing be possible? Most strategies for self-cleaning involve imparting the membrane's surfaces with chemical reactivity, in some cases activated by light or electricity. The generation of highly reactive species will break down many organic and biological materials into smaller molecular fragments, which will release from the surface and return to the permeate or retentate streams. If this approach can be perfected, it may be possible to install membrane-based water treatment systems that rarely, if ever, require replacement.

forward osmosis (FO)   reverse osmosis (RO)

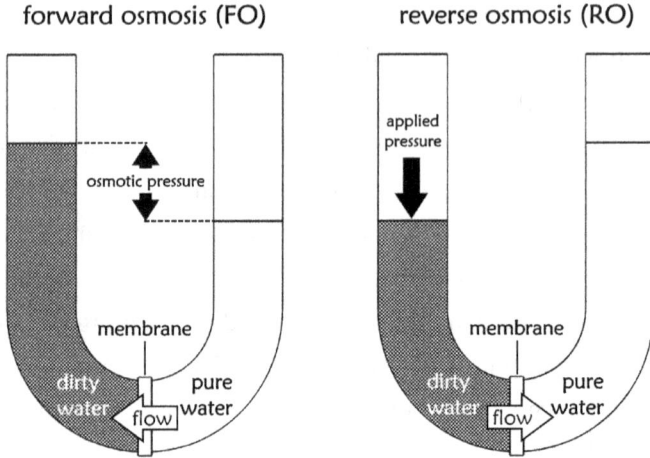

Figure 8.9.   Forward and reverse osmosis diagrams.

While RO has dominated the desalination world in recent years, its non-reverse cousin, osmosis (often called "forward osmosis," or FO, to distinguish it from RO), has caught the attention of researchers for some potential benefits in water treatment. Whereas in RO substantial pressure is applied to overpower the natural osmotic pressure across a semi-permeable membrane, FO exploits this osmotic pressure to accomplish a treatment process (Figure 8.9), thereby offering a route to save energy.

In order to use FO to treat water, you have to create a pressure gradient that will pull water through the membrane from the feed stream. To accomplish this feat, a "draw" solution is created on the permeate side of the membrane. This draw solution has a higher concentration of solutes than the feed solution, so osmotic pressure works to transport water from the feed side into the draw side. Clearly, then, the resulting (now diluted) draw solution is not "clean" water as you would produce in a typical RO process. However, if there is a simple, low-energy means of removing the solutes from the draw solution, clean water is achievable. One idea explored by researchers is to use tiny magnetic particles as the solute on the draw side, creating the osmotic pressure. After water has been pulled into that draw solution, the particles can be separated from the water by applying a strong magnetic field to the side of the container/pipe, sequestering all the particles against the wall. Typically, FO is used to treat highly contaminated water and make it more suitable for disposal or further treatment.

Size exclusion, the primary mode of operation for membrane-based separations, is not always the best or only solution to treat a particular stream of water. Some treatment processes can be accomplished using materials that are designed to exhibit an affinity to bind targeted pollutants. The most common such class of materials is called "sorbents." These are materials that absorb (with a "B") or adsorb (with a "D") other materials, by physically trapping them or through some sort of affinity based on chemistry or electrical charges. What, you ask, is the difference between absorption and adsorption? Absorption is the term most folks use in common parlance for soaking up a fluid, but this term is frequently misused. Absorption occurs when a fluid dissolves or permeates into another liquid or solid, and it has few applications when it comes to water — although superabsorbent polymers are an important exception with functions ranging from hygiene to waste solidification. Much more pertinent is adsorption, which takes place by adhesion of atoms, ions, or molecules onto a surface. Generally, an adsorbent is designed to be highly porous such that the internal surface area is large, accommodating many more adsorbates in a given volume. A kitchen sponge actually functions primarily by adsorption, with fluid filling the void spaces. By far the most widely used adsorbent for water treatment is activated carbon, which is highly porous. It is made by heating carbon sources in an oxygen-free environment and exposing them to an oxidizing agent like steam or carbon dioxide. This environment drives reactions that form three-dimensional microstructures within the carbon, with pore sizes tunable by reaction time. Activated carbon can adsorb many substances that end up in water.

In water treatment systems, there is an important subcategory of adsorption called "ion exchange." This process involves the swapping of ions between a water stream and a surface — usually a polymer resin. Common applications for ion exchange include water softening, purification, and decontamination. Softening is achieved by exchanging multivalent cations such as calcium (Ca^{2+}) and magnesium (Mg^{2+}) in the water for monovalent species such as sodium (Na^+), potassium (K^+), or protons (H^+). Harmful ions in water such as nitrates and nitrites can also be adsorbed using ion exchangers. Since the ion exchange process is consumptive, these systems have a limited lifetime, and the exchange medium must eventually be regenerated by flushing it with a highly concentrated solution of replacement ions. Because this regeneration process produces a lot of waste, applications in large-scale operations such as municipal wastewater treatment are challenging. Water softening is needed because the multivalent cations precipitate and form scale, essentially a white scum on sinks, tubs, and cooking and eating utensils. Water softening is typically required when a municipality or individual home uses "hard" well water.

The future of sorbents in water systems will rely on a collection of proven technologies such as activated carbon and ion exchange resins and novel, innovative sorbents capable of higher specificity or scalability. What if a sorbent were capable, for example, of selectively grabbing bioactive molecules such as endocrine disruptors from wastewater treatment plants while allowing non-hazardous organic materials to pass through? Another enticing prospect is sustainable (i.e. easy and non-consumptive) reusability, which is a potential game-changing property for sorbent materials as a dramatic cost-reducer. One means of reuse relevant for certain sorption applications is mechanical compression. There have been reports of sponges capable of selectively adsorbing oil from an oil-water mixture (think oil spills in bodies of water). Once oil has been extracted from the water, it can be squeezed out into a containment vessel, thereby emptying the pores of the sponge and rendering it ready for another cycle of use.

With either membranes or sorbents, you have to transport the entire stream of water through the porous material to remove contaminants, and that translates into a lot of energy expenditure. If the contaminant in question happens to be electrically charged, there are other, potentially more efficient options available. Using electric fields, a process can

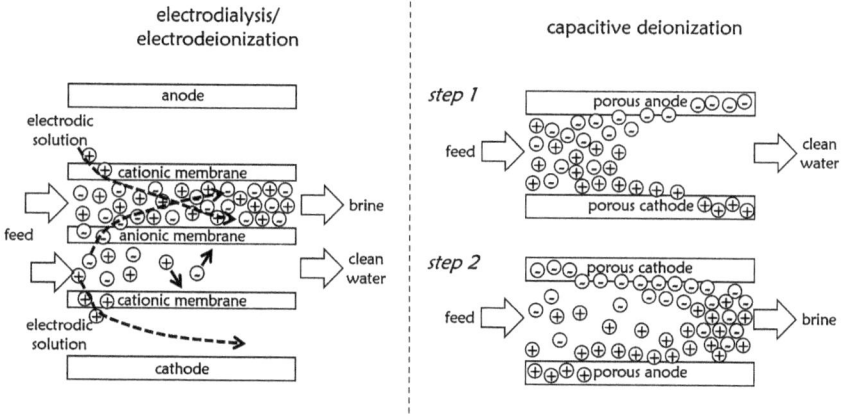

Figure 8.10.   Field-dependent separations techniques. In electrodialysis and electrodeionization, membranes that selectively transport either anions or cations are a central piece of the system. The brine is continuously generated. For capacitive deionization, the porous electrodes eventually become saturated with ions, at which point the direction of the electric field is reversed, releasing the ions and forming a concentrated brine for discharge.

focus on moving only the (charged) contaminants. This can be especially beneficial when the concentration of charged species is relatively low, such as with brackish groundwater. Electrodialysis, electrodeionization, and capacitive deionization are all examples of electric field-dependent separations (Figure 8.10). As these technologies continue to evolve, opportunities will likely emerge to selectively remove particular charged species, potentially in a resource-recovery system that captures valuable materials out of wastewater streams. Researchers are investigating the recovery of lithium for batteries or uranium for nuclear energy during seawater or brackish water desalination. Other researchers are proposing to operate electrically driven purification in both directions. Removing salt or ions from water requires energy, as discussed back in Chapter 1. In reverse, dissolving concentrated salts or ions into clean water will generate a small amount of energy. If you save some of the brine concentrate in storage tanks, at times when there is a power shortage, you can operate the system in reverse and generate electricity for a short period. If you are mixing the concentrated brine into brackish feed water, you will generate most of the energy but not waste any of the clean water. As we suggested earlier in this chapter, the future will include managing water and energy as an integrated system.

An outlook toward a sustainable water future

Water is the most important material. Humans have alternately prospered and suffered at the hands of water since pre-historic times. Water underlies world-changing developments for millennia, including the rise and fall of numerous civilizations. Without ample supplies of safe water, we cannot grow food to sustain ourselves, we cannot manufacture goods, and we cannot create energy to power society — it is not much of an exaggeration to say *water is ... everything*. Population growth, economic development, climate change, pollution, and urbanization are all working in concert to increase pressure on water resources. Most of the aquifers we rely on for routine water consumption are being pumped dry. Many major river systems are so heavily exploited they no longer even reach the sea in some years. Droughts, a plague on societies throughout our history, are becoming increasingly severe, disrupting civil order. Wetlands are steadily disappearing as a result of development and climate change. Those water supplies that do remain are increasingly polluted, requiring more expensive treatment and threatening health and environment. Thousands die every day because they lack access to clean water. Our challenges with water originate from a combination of ineffective management, neglect, and disregard for natural systems. Eilon Adar from Ben Gurion University of the Negev, Israel says "We don't have a water shortage problem, we have a water management problem." We use water wastefully, improperly incentivizing its use. We have not invested in water infrastructure, both in terms of upkeep of existing systems and in the implementation and expansion of new systems. Our mindset has been stubbornly centered on identifying ever more supplies of water to tap in an effort to meet an unquenchable thirst. We are learning.

Moving forward, we must boldly shift toward a new paradigm to ensure a sustainable water future. Simplest among the myriad changes necessary are simple water conservation and efficiency measures. The easiest and cheapest source of water is the one you don't use. Implementing proper incentives — especially in pricing — can be a powerful tool to achieve reductions in water use. Installing low-flow fixtures, repairing leaks, and the like are actions that all residential and business consumers can implement. Here, though, it is important to keep in mind that reducing the consumption of things that aren't water still has a direct impact on water consumption. Eating less meat, dairy, and highly processed foods and more locally produced food can have a dramatic impact. Recycling and reusing materials minimizes the need for additional manufacturing, which requires water. Turning off the lights and using more energy-efficient appliances lowers electricity demand and thereby water withdrawals for cooling in thermoelectric power plants. Essentially all activities that we already consider supportive of sustainability in other arenas also help push us toward better water sustainability.

At the municipal level, the plan is primarily to concentrate on infrastructure investments. In many regions, repairing and replacing water mains that have outlived their useful lifetime is essential to minimizing losses of treated water through widespread leaks. These activities provide a valuable opportunity to install additional networks of pipes to accommodate multiple streams of water intended for different purposes and, therefore, having different levels of purity. Many water utilities also need to upgrade their wastewater treatment facilities, both to reduce the level of pollutants leaving the plants and to recover useful resources from the waste such as phosphorous and organic matter.

Zooming out even further, the national and international communities must begin to properly oversee water resources. An area currently devoid of international oversight is the trade in virtual water. Movement of goods around the globe is heavily regulated, for sure, but considerations of impacts on regional water resources tied to this trade are absent. Surely we are not using our worldwide water resources in the most sustainable manner when private interests are given free rein to search the world for low-cost, but often limited water resources (such as from aquifers) for manufacturing goods intended for use in some disparate corner of the earth. We must find mechanisms to share these resources equitably and sustainably. No doubt we will need to consider massive shifts of water-intensive activities to geographic areas that are endowed with plentiful sources of sustainable water. We can wait until the groundwater is gone to take such action, and face grave, disruptive crises; or we can step back and identify where such shifts are needed now so they can occur in an orderly and equitable manner.

Arguably the most exciting direction for our water future is the development of entirely new water technologies. Water research is exciting from an international perspective. For example, while the United States and China may be hesitant to share cyber technologies, we can confidently share water technologies. Each nation will implement innovative solutions to help their own people without competing in global trade markets. Water systems of the future will be equipped with a vast array of sensors that can reliably detect a host of potential concerns, providing that information in real time to control systems that can alert engineers or homeowners that action is needed or even execute a treatment process to address the concern without human intervention. Leaks will be detected as soon as they occur so repairs can be scheduled before losses become extreme. New materials will be engineered to pull fresh water directly from the moisture in the air, even in arid regions. This will be a life-changing shift in the developing world, where so many people are trapped in the cycle of manually fetching water from a distant source for their daily needs. Components in water treatment systems will have far longer operational lifetimes thanks to self-cleaning capabilities. Water treatment systems will become increasingly efficient and effective, ultimately enabling the loop of water use to be closed, short-circuiting the natural water cycle so that withdrawals from traditional sources can be substantially curtailed. Effluent from wastewater treatment plants will be clean enough for direct reuse, including for human consumption. In these future systems, energy use to treat and convey water will plateau and eventually decline. A water system that stresses energy systems will not be sustainable.

Water crises are real, and they threaten to disrupt society on an unprecedented scale. But with proper water management, policy, and technology research and development, with a shift in our perspective from consumption to sustainability, with a recognition that access to clean water is a universal human right, we can meet this challenge and enable a secure water future for ourselves and for generations to come.

www.ingramcontent.com/pod-product-compliance
Lightning Source LLC
Chambersburg PA
CBHW061250220326
41599CB00028B/5601